# Proteotronics

# Proteotronics

## Development of Protein-Based Electronics

Eleonora Alfinito
Jeremy Pousset
Lino Reggiani

Pan Stanford Publishing

*Published by*

Pan Stanford Publishing Pte. Ltd.
Penthouse Level, Suntec Tower 3
8 Temasek Boulevard
Singapore 038988

Email: editorial@panstanford.com
Web: www.panstanford.com

**British Library Cataloguing-in-Publication Data**
A catalogue record for this book is available from the British Library.

**Proteotronics: Development of Protein-Based Electronics**



Cover Image: A futurist multi-odour smell-sensor based on olfactory receptors belonging to the family of G protein–coupled receptors (GPCRs).

ISBN 978-981-4613-63-7 (Hardcover)
ISBN 978-981-4613-64-4 (eBook)

Printed in the USA

# Contents

# Preface

Protein-mediated charge transport is of relevant importance in the design of protein-based electronics and in attaining an adequate level of understanding of protein functioning. This is particularly true for the case of transmembrane proteins, like those pertaining to the G protein–coupled receptors (GPCRs), which are involved in a broad range of biological processes, and a large number of clinically used drugs that elicit their biological effects via a GPCR.

This book aims to review a variety of experiments devoted to the investigation of charge transport in proteins and presents a unified theoretical model to interpret macroscopic results in terms of the amino acids backbone structure of the single protein. The book should serve a broad audience of researchers involved in the field of electrical characterization of biological materials and in the development of new molecular devices based on proteins, such as nanometric biological sensors of new generation. The book should also serve as a reference platform that surveys existing data and provides the basis for the future development of a new branch of nanoelectronics, which by mixing proteomics—that is, the large-scale study of proteins, particularly their structures and functions—and electronics is introduced here as *proteotronics*. The main objective of proteotronics is to propose and achieve innovative electronic devices, based on the selective action of specific proteins.

**Eleonora Alfinito**
**Jeremy Pousset**
**Lino Reggiani**
Summer 2015

# Chapter 1

# Introduction

## 1.1 General on Proteins

The word *protein* was used for the first time in 1838 by J.G. Mulder [Mulder (1839)] to describe the *oxidized basic radical* that, when combined in simple ratios with sulfur and phosphorus, should reproduce the same substances found in silk, egg albumin and serum, gelatin, and so on [Vickery (1950)]. Therefore, the name, from late Greek (proteios or primary), describes its basic role in chemistry [Branden (1991)].

Proteins appear like linear chains of lego-like building blocks (amino acids) able to swiftly (few microseconds [Kubelka (2004)]) organize in space so as encoded in this sequence. As a matter of fact, they are self-assembled macromolecules: they both fold their amino acids chain (intramolecular self-assembly) and also combine with other proteins to produce super-molecular structures (intermolecular self-assembly). Finally, proteins are the most simple, efficient, and irreplaceable molecular machines in the three domains of life: *Archaea, Bacteria, Eukarya* [Balch (1977); Woese et al. (1990)]. Proteins contribute about 50% to the biological matter that constitutes the three domains of life and are of basic relevance in almost all living processes. They are involved

*Proteotronics: Development of Protein-Based Electronics*
Eleonora Alfinito, Jeremy Pousset, and Lino Reggiani

ISBN 978-981-4613-63-7 (Hardcover), 978-981-4613-64-4 (eBook)
www.panstanford.com

practically in any biological process where their importance is widely diversified. Their functions span from catalysis in chemical reactions to maintenance of the electrochemical potential across the cell membrane, from transport of any kind of substance to control of gene functioning. In particular, one should mention:

(i) the exchange function of material between the different parts of the cell and the external environment (channels, membrane transport, etc.);
(ii) chemical connection among different cells (hormones, cytokines, membrane receptor, etc.);
(iii) immunity protection (antibody, hysto-compatibility of complexes, etc.);
(iv) energetic functions (breathing and photosynthetics complexes);
(v) mobile functions (actin, myosin, etc.); and
(vi) correct expression of genetic information.

## 1.2 Structural Properties

Proteins are classified as *simple*, when their constituents are only amino acids, and as *conjugated*, when they contain other chemical groups (*prosthetic groups*) such as lipids, oligosaccharide chains, metals, phosphoric acid, and pigments [Nelson and Michael (2004)]. The basic units underlying a protein are the amino acids. Only 20 amino acids in various combinations form proteins; so for simplicity, each amino acid is often represented by a letter of the alphabet. The 20 standard plus two additional and four ambiguous amino acids are reported with their symbols and polar properties in Table 1.1.

In other words, proteins are polymers constituted by 20 different types of amino acids (the alphabet of proteins) with a sequential length containing 20–1000 amino acids.

Amino acids share a common structure in which a carbon atom ($\alpha$-carbon, $C_\alpha$) is linked to an amine group ($-NH_2$) and to a carboxylic group ($-COOH$), which provide the optical activity to the amino acids. The central atom $C_\alpha$ is also linked to an $H$ atom and to a lateral chain, usually called residue and labelled as $R$, whose structure is

**Table 1.1** List of known amino acids

| Amino acid | 3-Letters | 1-Letter | Side-chain polarity | Polarizability ($Å^3$) |
|---|---|---|---|---|
| Alanine | Ala | A | nonpolar | 1.1 |
| Arginine | Arg | R | basic polar | 5.7 |
| Asparagine | Asn | N | polar | 9.8 |
| Aspartic acid | Asp | D | acidic polar | 4.0 |
| Cysteine | Cys | C | nonpolar | 1.8 |
| Glutamic acid | Glu | E | acidic polar | 6.2 |
| Glutamine | Gln | Q | polar | 17.7 |
| Glycine | Gly | G | nonpolar | 2.2 |
| Histidine | His | H | basic polar | 12.3 |
| Isoleucine | Ile | I | nonpolar | 1.1 |
| Leucine | Leu | L | nonpolar | 1.2 |
| Lysine | Lys | K | basic polar | 8.4 |
| Methionine | Met | M | nonpolar | 2.8 |
| Phenylalanine | Phe | F | nonpolar | 1.0 |
| Proline | Pro | P | nonpolar | 1.3 |
| Serine | Ser | S | polar | 7.3 |
| Threonine | Thr | T | polar | 8.2 |
| Tryptophan | Trp | W | nonpolar | 1.8 |
| Tyrosine | Tyr | Y | polar | 5.3 |
| Valine | Val | V | nonpolar | 1.0 |
| Selenocysteine | Sec | U | undefined | undefined |
| Pyrrolysine | Pyl | O | undefined | undefined |
| Asparagine or aspartic acid | Asx | B | undefined | undefined |
| Glutamine or glutamic acid | Glx | Z | undefined | undefined |
| Leucine or isoleucine | Xle | J | undefined | undefined |
| Unspecified or unknown | Xaa | X | undefined | undefined |

a specific signature of each amino acid. The first amino acid in the chain retains its free amino group, and this end is usually called the *N-terminus* of the protein. Analogously, the last amino acid to be added is left with its free carboxy group, and this end is often called *C-terminus*.

Figure 1.1 depicts the typical structure of a given amino acid as described above. The three-dimensional (3D or tertiary) structure of a protein is the result of the creation of peptide bonds among amino acids. A peptide bond is a covalent bond linking the carboxylic group of one amino acid to the amine group of another amino

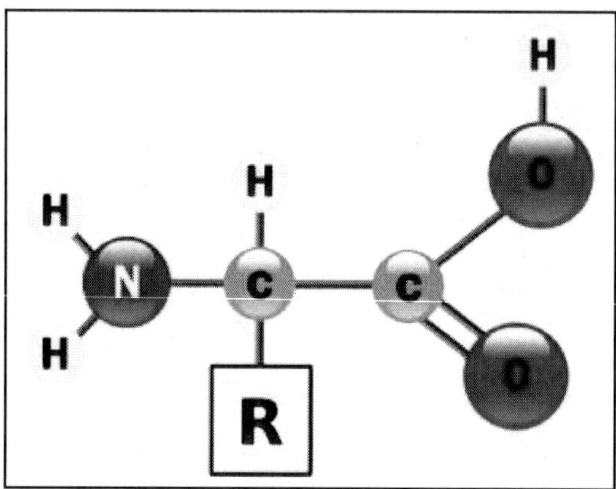


**Figure 1.1** Schematic of the atomic structure of amino acids with $R$ denoting the residue. Only 20 amino acids, in various combinations, form proteins. For simplicity, each amino acid is often represented by a letter of the alphabet. The 20 standard plus two additional and four ambiguous amino acids are reported with their symbols and polar properties in Table 1.1 (Image courtesy: Yassine Mrabet).

acid. When many amino acids are linked by peptide bonds, the resulting molecule is called a polypeptide, from which the name of the polypeptide chain is usually taken as synonymous of protein. More precisely, the term polypeptide is referred to as a structural unity consisting only of amino acids. By contrast, a protein is a functional unity consisting of a single or many polypeptides that are folded in appropriate three-dimensional configurations and able to develop particular functions. The atoms stemming in the peptide bond constitute the main chain of the protein, also called the backbone, while other atoms constitute the lateral chains. Once formed, proteins tend to fold on themselves owing to some weak bonds (hydrogen bond) or to stronger bonds (disulfide bond). The rigid bond belonging to amino acids determines the conformational structure of the protein, while the rotational degrees of freedom of the peptide bonds lead to a very flexible backbone [Reuveni et al. (2008)]. The particular conformation adopted by a peptide chain is determined by rotations around C–C and C–N amide bonds and the resulting dihedral angles, known as the $\psi$ and $\phi$ angles, respectively. These angles are named after the Indian biophysicist

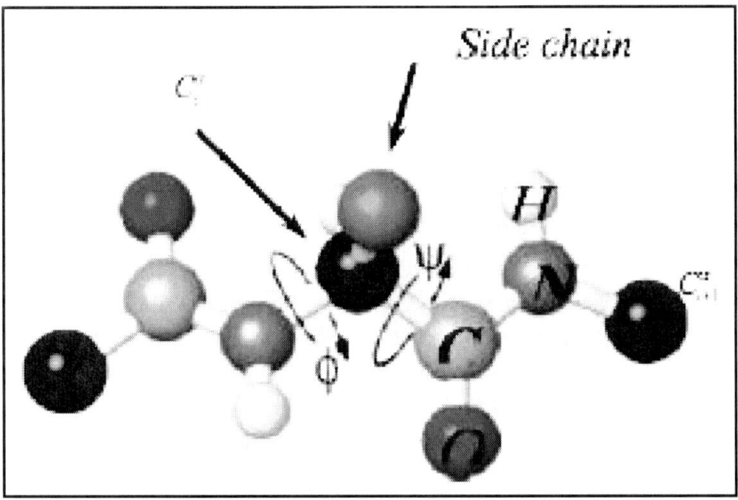


**Figure 1.2** Schematic of a peptide group.

G. N. Ramachandran, who began studying them in the 1960s [Ramachandran et al. (1963)]. Overall, due to the complexity of the structure, a protein can be considered a polymer or more properly an heteropolymer.

Figure 1.2 reports two peptide groups with the indication of the rotational degrees of freedom (dihedral angles $\phi$ and $\psi$) that provide the backbone with the given flexibility.

Quaternary, tertiary, and secondary structures, but not primary structures, can be disrupted: This protein state is called denatured state and occurs at high temperatures or in the presence of very aggressive solvents. Since we are in the presence of a disordered structure, the denatured protein is not functionally active from a biological point of view. Most proteins also take some ordered configurations, and this more favorable configuration, in the absence of the specific ligand, is called the native state. The native state is the stable and functional tertiary structure, which is characterized by a minimum of the potential energy and by that unique conformation that allows the protein to carry out the proper function adequately. Often biochemists manipulate proteins in ice baths to keep a sufficient low temperature and, as a consequence, to preserve the active state. The process that forms the biosynthesis of the peptide leads to a protein structure in its native state, biologically active, which is called folding and will be detailed in next sections.

## 1.3 Structure Levels

The protein capability to interact with the host organism, more precisely to develop particular biologic functions, is strictly related to its tertiary structure. Indeed, this structure allows to exploit specific functions, such as those pertaining to control or enzymes, energy or hormones, since it promotes interaction among different proteins, with the DNA and/or with substances of different chemical nature.

The tertiary structure only depends upon the amino acids chain, which constitutes the so-called primary or covalent structure: Meanwhile, the interactions among amino acids determine the ordered conformation of the protein. This is known as the Anfinsen thermodynamics hypothesis [Anfinsen (1973); Anfinsen et al. (1961)]. The protein structure can be considered at different organization levels. To this end, the following four levels can be individuated:

(i) Primary structure: It corresponds to the specific sequence of the amino acids pertaining to the backbone. The polypeptide primary structures are determined by genes.
(ii) Secondary structure: The amino acids sequence of a polypeptide follows the physical-chemical laws and induces a polypeptide to fold in a more compact structure. The amino acids can rotate around the bonds that are present in a protein. As a consequence, proteins are flexible and can fold in a given number of forms, among which we recall $\alpha$ helices and $\beta$ sheets, also called reasons of secondary structures. The identification of the secondary structure of a given protein implies the determination of the type and the position of these sub-structures.
(iii) Tertiary structure: As anticipated, it is the global three-dimensional conformation of a single polypeptide. It corre-

sponds to the structure taken by the protein when in the native state or in one of the activated states. However, most of the proteins are composed of two or more polypeptides, thus showing a quaternary structure.

(iv) Quaternary structure: In general, it concerns with very large proteins (e.g., hemoglobin and acetylcholine) formed by two or more polypeptides, each of them presenting a tertiary structure and which assemble each other. Indeed, often these proteins consist of several units essentially equal to each other. In practice, the quaternary structure concerns with the spatial, topological, and functional disposition of these sub-unities [Spinozzi and Beltramini (2012)].

Figure 1.3 reports a schematic of the hierarchy structure of a protein. The structural levels of a protein discussed above are interdependent. Indeed, if a level changes, the others can change in consequence. For example, when the primary structure is modified, this modification influences the secondary, tertiary, and quaternary structures.

Figure 1.4 reports the schematic of an $\alpha$ helix and a $\beta$-plated sheet. In $\alpha$ helices, the protein backbone forms a regular helix, which in most cases is right-handed. As illustrated in Fig. 1.4, a hydrogen atom bonded to a nitrogen atom forms a hydrogen bond with an oxygen atom, which in turn is linked through a double bond to a carbon atom. These hydrogen bonds are found at regular intervals and induce the polypeptide backbone to form a helix.

By contrast, in the $\beta$ sheets, the regions of the polypeptide backbone are disposed parallel to each other. When these parallel regions form hydrogen bonds among the linked hydrogen to a nitrogen atom and an oxygen one involved in a double bond, the polypeptide chain takes a repeated zig-zag shape, also called $\beta$ many-folded plane, as evident in part (b) of Fig. 1.4. The $\alpha$ helices and the $\beta$ sheets are structures that determine the characteristics of a protein; for example, in some proteins, the $\alpha$ helices mostly consist of nonpolar amino acids. Proteins that contain many of these parallel regions with an $\alpha$ helix structure exhibit the tendency to anchor themselves in an environment rich in lipids, for example the plasmatic membrane of a cell [Andersen and Koeppe (2007)].

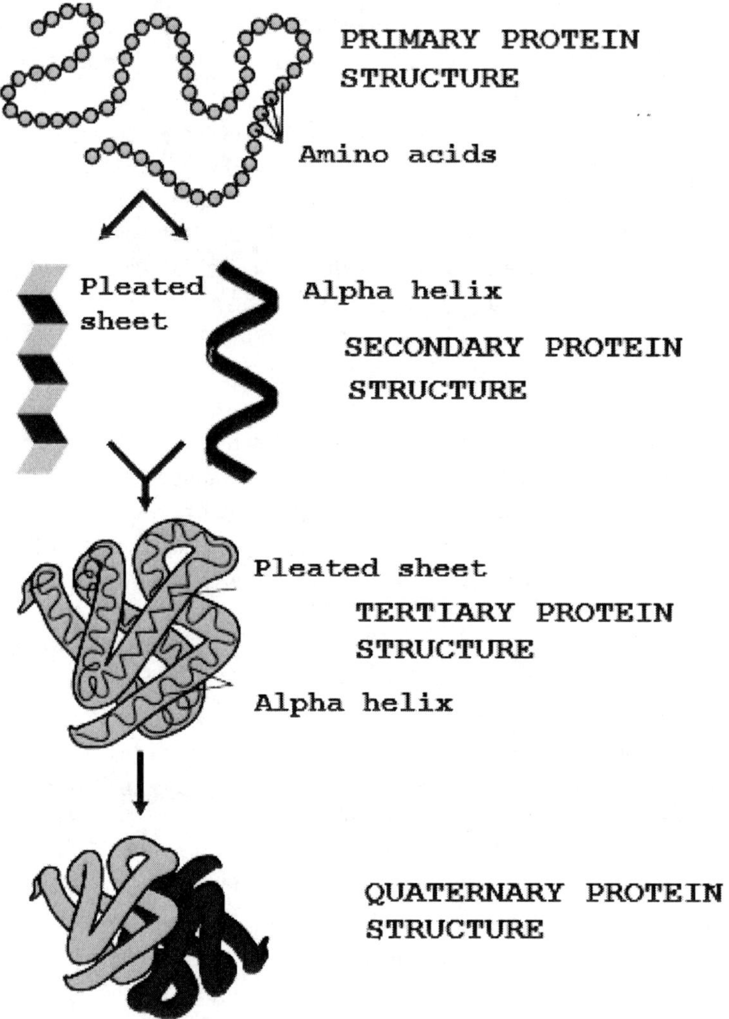


**Figure 1.3** The four-level hierarchy structure of a protein (Image courtesy: National Human Genome Research Institute).

The regions along the polypeptide chain that do not take an $\alpha$ helix conformation or a $\beta$-pleated sheet are called tangle regions, since they do not take a secondary structure.

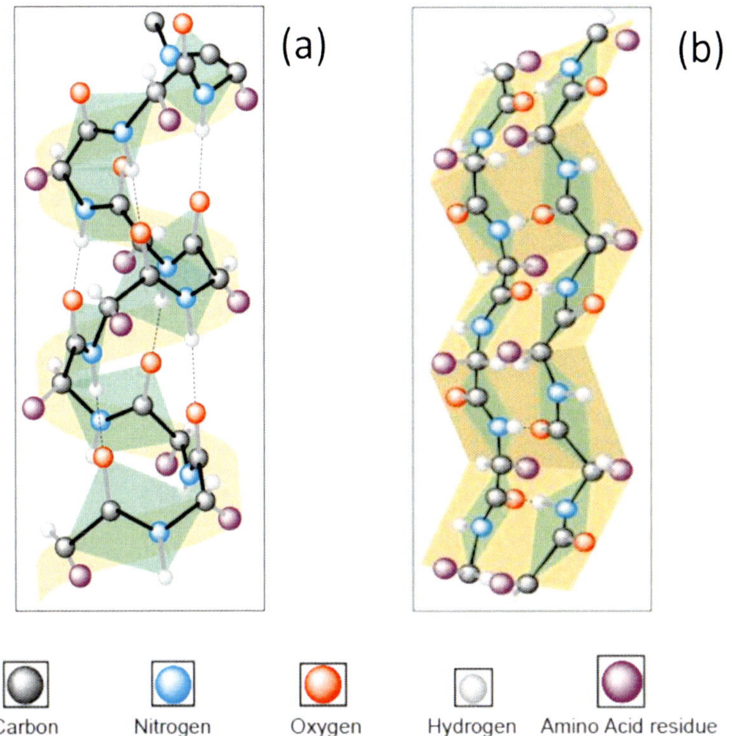


**Figure 1.4** Schematic of $\alpha$-helix (a) and $\beta$-pleated sheet (b).

## 1.4 Protein Folding

The folding of a protein consists of a complex phenomenon through which the amino acids chain finally achieves a three-dimensional stable shape [Anfinsen (1973); Echenique (2007)]. Figure 1.5 reports a schematic illustration of an unfolded protein and what it looks like once it has been folded. In the absence of the specific ligand, this structure is also called the native state. Millions of years of evolution in nature has determined a strong selection in the sequence of amino acids typical of each protein and, as a consequence, proteins exhibit quite peculiar properties, especially in relation to folding [Balbín and Andrade (2004)]. One should notice that despite their complexity, real proteins exhibit a well-

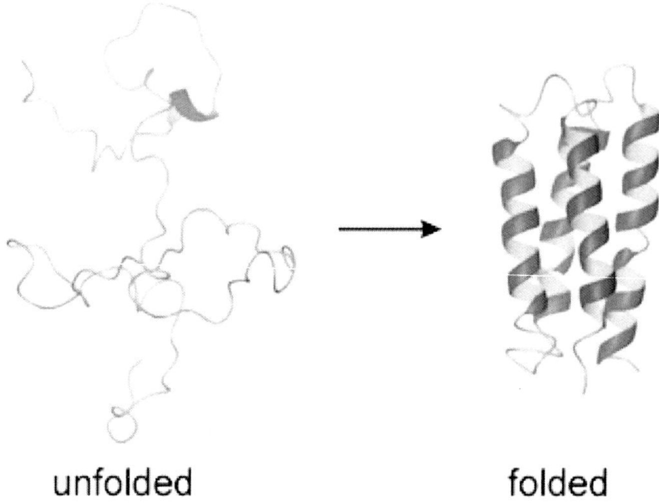


**Figure 1.5** Schematic of protein transition from an unfolded to a folded state (Image courtesy: Adam Liwo/Cornell University).

defined native state and fold themselves in this state through a fast and safe process. This means that a folding process is, in general, a reversible one. The correspondence between the amino acids sequence that forms the protein and its native state is often expressed by the sentence that *the native state of a protein is already codified in its primary structure.*

Among the reasons motivating the study of protein folding, the following two are most relevant. The first one faces the problem to formulate methods and models able to predict the native structure of a given protein starting from its amino acids sequence [Carloni et al. (2002); Fiser and Sali (2003); Go (1983); Hall et al. (2004); Kitao and Go (1999); Launay et al. (2012a,b); Roy et al. (2010); Vaidehi et al. (2002)]. The solution to this problem would enable the elaboration of techniques suited to provide realistic three-dimensional structures and thus to develop new proteins. The second reason is to understand the physical phenomena that are at the basis of the folding processes. In this direction, one should cite researches based on first principle approaches, see [Carloni et al. (2002); Kitao and Go (1999)], as well as the use of models based on a mean field approach, see [Cai and Zhou (2011); Gardino et al. (2009); Kobilka and Deupi (2007); Miyashita et al. (2005); Okazaki and Takada (2008); Park et al. (2008)], and effective algorithms for the prediction of three-dimensional structures [Carloni et al. (2002); Fiser and Sali (2003,?); Go (1983); Hall et al. (2004); Kitao and Go (1999); Launay et al. (2012a,b); Roy et al. (2010); Sali and Blundell (1990); Sutto et al. (2006); Vaidehi et al. (2002)].

Overall, the folding process remains a very complex phenomenon, and its understanding needs a strong interaction between experiment and theory. In essence, it is a multidisciplinary task (involving biologists, chemists, informatics, physicists, nanotechnologists, etc.) and also a multiscale task (folding involves structural movements covering time and spatial scales, which range over many orders of magnitudes). To date a single theoretical model able to describe such wide ranges of time and spatial scales is not available. Rather, one can rely on a multitude of different approaches finalized to describe limited cases of specific interest. To this purposes, first principle approaches have the drawback of being applicable to a limited range of spatial scales. More simple approaches, based on lattice models, can avoid this limitation. In fact, research on this problem is still in its infancy but rapidly developing.

## 1.5 Experimental Techniques to Investigate Structure and Functions of Proteins

The determination of the structure of a macromolecule, such as a protein, is a quite complex and elaborated process for which no predefined techniques are completely adequate. The knowledge of the primary structure is a prerequisite for most of the physical study, even if the determination of structure is essentially a biochemical task. In the last few decades, several projects contributed to provide the sequential chain of amino acids constituting any given protein; among them one should recall the human genome project [DeLisi (1988)], the consensus coding sequence, and the reference sequence project [Pruitt et al. (2012)].

To get information on the secondary structure of a given protein, several spectroscopic methods can be used, such as optical absorbance, linear and circular dichroism [Greenfield and Fasman (1969); Provencher and Gloeckner (1981); Rodger and Nordén (1997)], infrared diffusion [Barth (2007)], Fourier transform infrared spectroscopy [Jackson and Mantsch (1995)], Raman spectroscopy [Chi et al. (1998)]; for a review, see [Pelton and McLean (2000)]. In a typical spectroscopic experiment, an electromagnetic radiation of given wavelength strikes on a sample, thus perturbing the electrical and magnetic properties of the protein, which are detected by the corresponding response. Accordingly, nuclear magnetic resonance (NMR) [Wishart et al. (1992); Wuthrich (1986)] and electron paramagnetic resonance (EPR) [Weil and Bolton (2006)] can produce a rather detailed spectrum of the single residue of particular secondary structures of the protein under test. The NMR technique measures the absorption spectrum in the radio-frequency range of molecules in the presence of a strong magnetic field. These radiations induce nuclear spin transitions in particular atoms, thus providing information on the molecular structure. The EPR technique enables the detection and the analysis of chemical species that contain one or more unpaired electrons (paramagnetic species). These species include free radicals, ions of transition metals, defects in crystals, molecules in state of fundamental electron triplet (for example molecular oxygen) or induced by photo-excitation.

The protein tertiary structure can be characterized by X-ray crystallography and electron diffraction [Rhodes (2006)]. Both these methods require that macromolecules are assembled in a well-ordered crystalline matrix to produce diffraction patterns; however, this assembly is not always achievable for any protein of interest. In any case, the X-ray diffraction pattern of samples with a minor order degree is still able to provide structure information, such as subunity organization, typology, entity and orientation of the secondary structure, and typology and structure of complexes like those ligand-molecules. In the absence of a crystalline structure, electron microscopy is a useful method to provide relevant information on protein topology through high-resolution images, which provide evidence of protein localization and interaction between proteins.

Determining the correlation between the tertiary structure and the function of a given protein is actually the major challenge of structural biology. Accordingly, among the different techniques of physical-chemical investigations listed above, NMR spectroscopy has proved in recent decades to compete with the crystallographic approach in the determination of complex biopolymers such as proteins, nucleic acids, and polysaccharides. These results have become possible owing to the fast development in the fields of electronics, superconductor technology, and NMR theory. The use of NMR is fundamental since NMR offers the possibility to characterize the structure of biological macromolecules with a resolution at an atomic level under conditions very close to physiological ones, thus allowing the investigation of the internal dynamics of proteins, conformational equilibria, and interaction among proteins. To this purpose, a significant advance was made by Kurt Wüthrich, Nobel Prize winner for chemistry in 2002 for his contribution to the development of NMR spectroscopy finalized to the determination of the tertiary structure of biological macromolecules in solution. The NMR technique combines the following four main elements for the determination of protein structure. (i) The nuclear Overhauser effect [Overhauser (1953)] provides the necessary information to determine the global folding of the polymeric chain. It consists of a polarization effect of the nuclear spin observed in NMR spectroscopy. (ii) The assignment of the specific sequence to several hundreds of NMR peaks obtained from a protein. (iii) Instruments of structural computation for the interpretation of NMR data and the evaluation of the obtained molecular structures. (iv) NMR multidimensional techniques for an effective collection of data.

Figure 1.6 reports the first protein structure determined by the NMR technique. In their pioneer work during 1976–80, Richard Ernst (Nobel Prize winner for chemistry in 1991) and Kurt Wüthrich developed two-dimensional NMR techniques for applications to biologic macromolecules. Then in 1981, they regularly used H NMR two-dimensional experiments on a group of four homonuclear molecules with the objective to determine the protein structure. However, only in 1984, the resolution of the structure of the first protein (an inhibitor enzyme of the seminal protein of a bull) was successfully obtained by NMR.

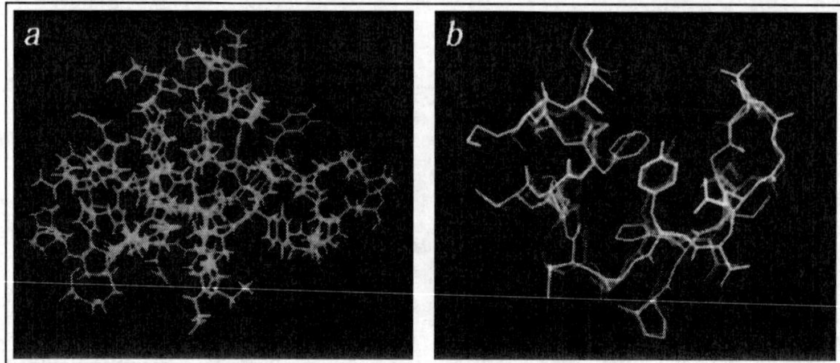


**Figure 1.6** First determination of a protein structure by NMR. Reprinted with permission from Macmillan Publishers Ltd. (*Nature Structural Biology*, doi:10.1038/nsb1101-923), copyright © 2001.

Recent scientific advances confirm that the systematic use of many complementary techniques accounting for their strength and weakness is still the best way to obtain a complete and detailed knowledge of the structure and physiological functioning of biological systems.

## 1.6 Classification of Proteins

Based on their shape and solubility, proteins can be broadly classified into three groups, which correlate with typical tertiary structures, namely, globular proteins, fibrous proteins, and membrane proteins.

Globular proteins are spherical (globe-like) proteins [Karplus et al. (1981); Pace and Hermans (1975)]. Figure 1.7 reports the ribbon representation of hemoglobin, a typical globular protein. Unlike the fibrous or membrane proteins, almost all globular proteins are soluble and many are enzymes. The term globin can refer more specifically to proteins, including the globin fold [Kendrew (1963)]. The spherical structure is induced by the protein tertiary structure. The apolar (hydrophobic) amino acids are bound toward the molecule interior, whereas polar (hydrophilic) amino acids are bound outward, allowing dipole–dipole interactions with the solvent, which explains the molecule solubility. A globular

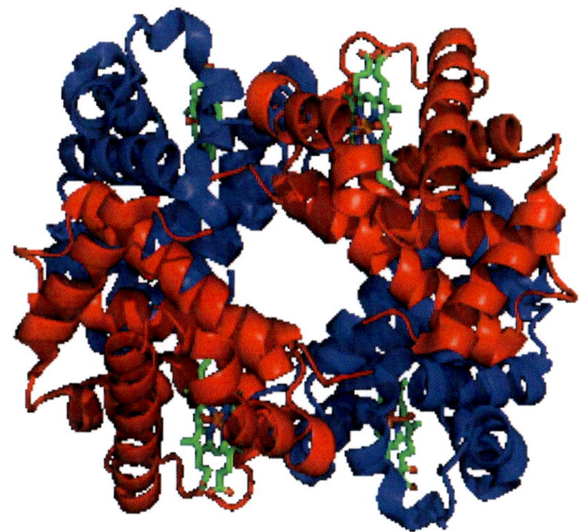

**Figure 1.7** Three-dimensional structure of hemoglobin, a globular protein (Image courtesy: Richard Wheeler).

protein is only marginally stable because the free energy released when the protein folded into its native conformation is relatively small. This is because protein folding requires entropy cost. Since a primary sequence of a polypeptide chain can form numerous conformations, a native globular structure restricts its conformation to a few only. It leads to a decrease in randomness, although noncovalent interactions, such as hydrophobic interactions, stabilize the structure. Globular proteins seem to have two mechanisms for protein folding, either the diffusion-collision model [Karplus and Weaver (1994)] or the nucleation condensation model [Fersht (1997); Itzhaki (1995)], although recent findings have shown globular proteins, such as PTP-BL PDZ2, that fold with characteristic features of both models. These new findings have shown that the transition states of proteins may affect the way they fold. These studies have shown that the folding of globular proteins affects their function [Bucciantini et al. (2002); Chiti and Dobson (2006)]. Globular proteins can act as:

(i) Enzymes, by catalyzing organic reactions taking place in the organism in mild conditions and with a great specificity. Different esterases fulfill this role.

(ii) Messengers, by transmitting messages to regulate biological processes. This function is accomplished by hormones, such as insulin.
(iii) Transporters of other molecules through membranes.
(iv) Stocks of amino acids.

Regulatory roles are also performed by globular proteins rather than fibrous proteins. Structural proteins, such as actin and tubulin, are globular and soluble as monomers, but polymerize to form long, stiff fibers. Among the most known globular proteins is hemoglobin, a member of the globin protein family. Other globular proteins are immunoglobulins (IgA, IgD, IgE, IgG, and IgM) and $\alpha$, $\beta$, and $\gamma$ globulins.

Fibrous proteins are often structural, such as collagen, the major component of connective tissue, or keratin, the protein component of hairs and nails. Fibrous proteins form rod- or wire-like shapes and are usually inert structural or storage proteins. They are generally water insoluble.

Figure 1.8 reports the ribbon representation of tropocollagen, a typical fibrous protein. Fibrous proteins are generally used to construct connective tissues, tendons, bone matrix, and muscle fiber. Examples of fibrous proteins include keratins, collagens, and elastins.

Membrane proteins are attached to, or associated with, the membrane of a cell or an organelle. These proteins are specifically targeted to different types of biological membranes [Andersen and Koeppe (2007)]. They are also the target of over 50% of all modern medicinal drugs [Dorsam and Gutkind (2007); Milligan (2004b)]. It is estimated that 20–30% of all genes in most genomes encode membrane proteins [Buck and Axel (1991); Wolfsberg et al. (1995)].

Figure 1.9 reports the ribbon representation of bovine rhodopsin, a typical transmembrane protein. Membrane proteins perform a variety of functions vital to the survival of organisms [Almen (2009); Bezanilla (2008); Greengard (1976); Yellen (2002)]. Membrane proteins can act as receptors by relaying signals between the cell internal and external environments. Transport proteins move molecules and ions across the membrane. They can be categorized according to the transporter classification database

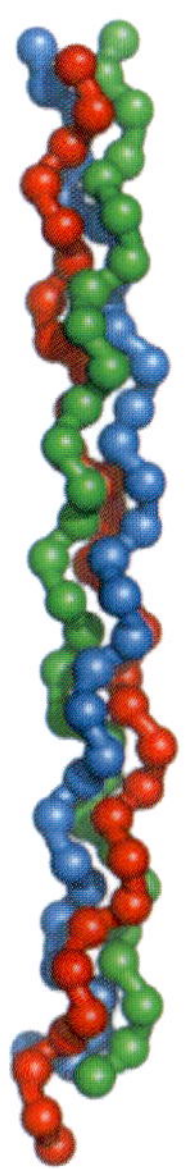

**Figure 1.8** Three-dimensional structure of tropocollagen triple helix, a fibrous protein.

membrane enzymes, for example oxidoreductases, transferases, and hydrolases. Cell adhesion molecules allow cells to identify each other and interact, for example proteins involved in immune response. Membrane proteins often serve as receptors or provide channels for polar or charged molecules to pass through the cell membrane [Friedrich et al. (2002); Lozier et al. (1975); Reeves (2008); Sato et al. (2008); Yellen (2002)]. Olfactory receptors (ORs) [Buck and Axel (1991); Malnic et al. (1999); Zozulya et al. (2001)] are transmembrane proteins belonging to the larger family of G protein-coupled receptors [Giraldo and Pin (2011); Kobilka and Deupi (2007); Lefkowitz (2004)]. In mammals, they control the smell-sensing action, and because of their excellent selectivity and sensitivity, they represent the gold standard for the realization of a smell nanobiosensor. Figure 1.10 reports the ribbon representation of bacterio rhodopsin (bR), a typical transmembrane protein belonging to the family of opsins, which, contrary to GRPRs, functions as a light receptor based on a proton pump mechanism.

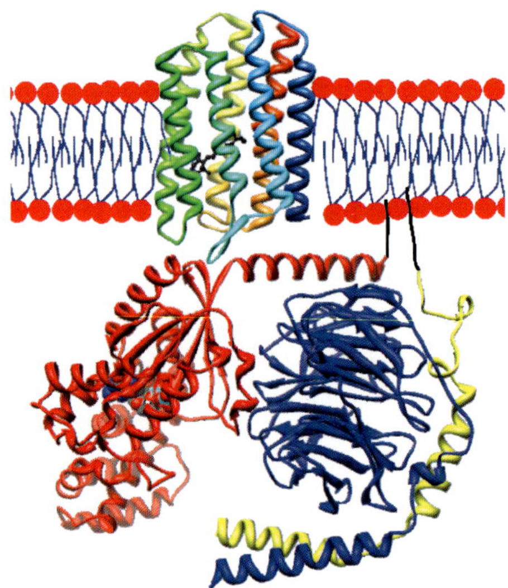

**Figure 1.9** Three-dimensional structure of sensory rhodopsin II (1gue.pdb), a transmembrane protein, embedded in the membrane with transducin (1got.pdb) under it. Rhodopsin is colored in a rainbow with the *N*-terminus red and the *C*-terminus blue. There is a bound retinal on the inside, which is colored black for ease of visualization. For the transducin, the Gt-$\alpha$ subunit is red, $\beta$ is blue, and $\gamma$ is yellow. Pseudo-anchoring sites are drawn in black. The Gt-$\alpha$ subunit has a bound GDP colored by atom (Image courtesy: Dpryan).

The sense of olfaction in mammals is the most complex and archaic: It is fundamental for identifying prey and predators and, in general, for avoiding toxic food. Recently, this efficient mechanism has been considered useful for the realization of a new concept of smell sensor, as proposed by the Single PrOTein Nanobiosensor Grid Array (SPOT-NOSED) and the bioelectronic olfactory neuron device (BOND) projects [BOND (2009–2011); SPOT-NOSED (2003–2005)]. Conventional smell sensors, usually called *electronic noses*, have been produced since several years [Comini (2002); Matteo Pardo (2009); Quaranta et al. (1999); Vo-Dinh et al. (2001)]. They usually perform the detection by means of a solid-state sensitive part, translating the capture information into an electrical signal. The signal is then digitized and elaborated by appropriate algorithms. At present, the sensitivity and selectivity of these conventional

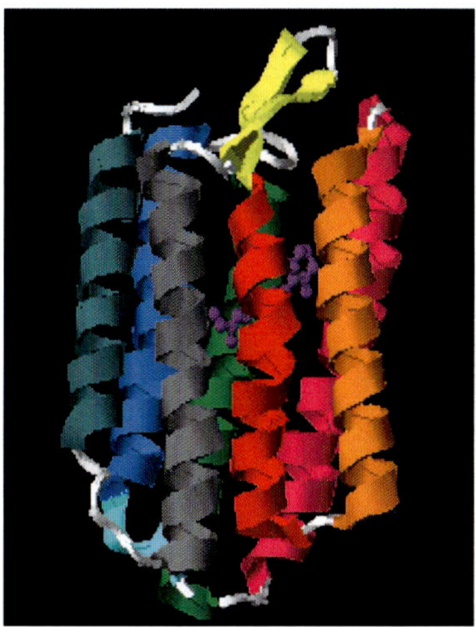

**Figure 1.10** Three-dimensional structure of sensory bacterio rhodopsin single monomer, a transmembrane protein, with retinal molecule between seven vertical $\alpha$-helixes. One more small helix is light blue, a $\beta$ sheet is yellow (Image courtesy: Nishikawa et al. 2005).

devices are high but still very far from those of the mammal smell-sense. To improve detector features, many efforts are at present devoted to the realization of biomimetic smell sensors, able to integrate the sensitivity and specificity of ORs with leader and user-friendly electronic devices [BOND (2009–2011); Gardner (1991); Göpel et al. (1998); Jianrong et al. (2004); Kim et al. (2009); Lee et al. (2012, 2009); Liu et al. (2006); Matteo Pardo (2009); Mombaerts (1996, 2004); Park et al. (2012); SPOT-NOSED (2003–2005); Wu et al. (2009, 2012); Yoon et al. (2009)]. The main objective is to produce a high selective sensor, potentially sensitive to all the possible odorants. Of course, it is not conceivable to reproduce the complexity of the in vivo olfaction system into such a device; therefore, it has been necessary to identify a specific ring of the olfaction detection chain, sufficient to give the notice of capture. The selected ring is the conformational change the ORs undergo when they bind one of their specific odorant molecules (ligands). It has been conjectured that the conformational change is already sufficient to modify the electrical properties of a protein appropriately anchored between Ohmic contacts [BOND (2009–2011); SPOT-NOSED (2003–2005)]. The validation of this conjecture disclosed the possibility to realize a nanobiosensor paralleling what the smelling sense is doing in nature.

# Chapter 2

# Sensing Proteins

Sensing proteins (receptors), to which specific signaling molecules may attach, are embedded in the cell membrane (cell surface receptors), or in the cytoplasm, or in the cell nucleus (nuclear receptors) [Gether and Kobilka (1998); Ghanouni et al. (2001); Kobilka and Deupi (2007)]. A molecule that binds to a receptor is called a ligand and can be a peptide (short protein) or another small molecule such as neurotransmitters, hormones, pharmaceutical drugs, or toxins. A particular family of receptors is that of opsins, which are sensitive to light, that is, photons [Bergo et al. (2004); Dioumaev et al. (2002); Friedrich et al. (2002); Lozier et al. (1975); Luecke (2000); Standfuss (2011)].

Numerous types of receptors are found in a typical cell. Each type is linked to a specific biochemical pathway and binds only to given ligand shapes, similar to how locks require specifically shaped keys to open. When a ligand binds to its corresponding receptor, it activates or inhibits the biochemical pathway associated with the receptor.

Ligand binding (or photon absorption in case of light) changes the conformation (tertiary structure) of the receptor molecule. This alters the space arrangement of different parts of the protein, changing the interaction of the receptor molecule with the host cell,

*Proteotronics: Development of Protein-Based Electronics*
Eleonora Alfinito, Jeremy Pousset, and Lino Reggiani

ISBN 978-981-4613-63-7 (Hardcover), 978-981-4613-64-4 (eBook)
www.panstanford.com

leading in turn to a cascade of events mediated by the associated biochemical pathway. Ligands that produce a protein response opposite to that of common ligands are called inverse ligands [Leff (1995)]. However, some ligands, called antagonists, merely block receptors from binding to other ligands without inducing any response themselves.

In recent years, the seven-helix transmembrane receptors sensitive to light (opsins family) and odor (OR family) are distinguished among the class of membrane proteins as very promising for applied purposes in the field of cellular pharmaceutics, sensors, etc. An important achievement is the 2012 Nobel Prize for Chemistry awarded to Brian Kobilka (Stanford) and Robert Lefkowitz (Duke) for their work on G protein–coupled receptors (GPCRs). Figure 2.1 reports a schematic drawing of a GPCR.

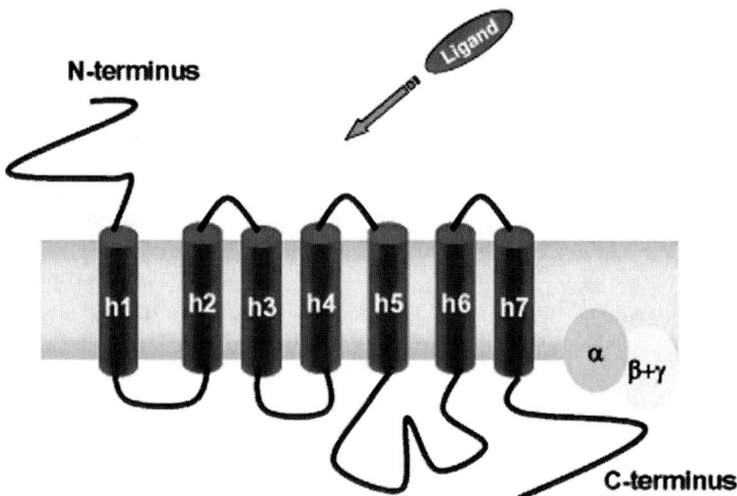


**Figure 2.1** Schematic drawing of a GPCR. The rectangular box represents the cellular membrane that separates the extracellular region (above) from the cytoplasmic region (below). The seven cylinders show the seven transmembrane $\alpha$-helical domains of the receptor, while the loops and the two termini ($N$-terminus and $C$-terminus) are shown by black lines. The two ellipses on the bottom represent the $\gamma$ and $\alpha + \beta$ subunits of the G protein and, finally, the ellipse on the top shows a ligand captured by the receptor.

This chapter briefly reviews the main mechanism of sensing and the set of electrical measurements to exploit the possibility to transfer the sensing action (in particular, the conformational change) into a change of an electrical property of the structure (device) to which the protein is more or less directly anchored. The correlation between sensing action and change of an electrical signal lays the groundwork for building a new generation of sensors based on biological materials, the so-called nanobiosensors.

## 2.1 Type-One Opsins

A prototype of these light receptors is bacterio rhodopsin (bR), which is an integral membrane protein commonly found in the so-called purple membrane (PM) of the *Archeobacterium salinarum*. PM contains only this protein that gives color to the membrane [Lozier et al. (1975)] and some lipids [Corcelli (2002)]. It is the energy harvester of the cell and can occupy up to nearly 50% of the whole surface area [Lozier et al. (1975); Luecke (2000)]. It appears as a two-dimensional hexagonal lattice whose unit cell is constituted by three identical proteins, each rotated by 120 degrees relative to each other, a trimer. Each protein consists of seven transmembrane $\alpha$ helices and contains one molecule of retinal buried deep within the typical structure for retinylidene proteins. It is the retinal molecule that changes its conformation when absorbing a photon, resulting in a conformational change of the surrounding protein and the proton-pumping action [Luecke et al. (1999)]. It is covalently linked to Lys216 in the chromophore by Schiff base action. The absorption of photons induces the photoisomerization of the retinal from the *all-trans* to the *13-cis* configuration. The photoisomerization activates a chain of events with a proton transfer from the protein to the extracellular side of the membrane. A thermal re-isomerization of the retinal restores the initial configuration [Pebay-Peyroula (1997)].

Another type-one opsin that acts similar to bR is proteorhodopsin (pR), which is found in different marine bacteria but also in archaea and eukaryotes [Bergo et al. (2004); Dioumaev et al. (2002); Reckel et al. (2011)].

## 2.2 G Protein–Coupled Receptors

As previously recalled, opsins of type two, typically bovine rhodopsin (BR) [Palczewski et al. (2000)], like olfactory receptors (ORs), belong to the huge family of GPCRs [Lefkowitz (2000)], proteins devoted to infer information from the environment (the presence of light, odors, pheromones, toxins, etc.) and to transmit it to the nervous system. A GPCR is activated by an external signal in the form of a ligand or other signal mediator. This creates a conformational change in the receptor, with rotation and tilting of some helices [Lefkowitz (2000, 2004)]. Furthermore, the ligand capture gives the G protein the activation signal. The specific action of G protein depends on the type of G protein. In particular, when an effector enzyme, or second messenger [Lodish et al. (2000)], is present in the cellular membrane, the G protein undergoes an activation/deactivation cycle, thus passing from a inactivated state linked to a guanosine diphosphate (GDP) to an activated state linked to a guanosine triphosphate (GTP), which gives it the function of a control molecular switch [Lodish et al. (2000)].

Figure 2.2 reports the detecting scheme of a GPCR, which, in the following, is detailed in what concerns the activation cycle of the trimeric G proteins. In the inactivated state, the subunit $\alpha$ is linked to the GDP and to the $\beta\gamma$ complex (Fig. 2.2, step 1). In this configuration, the G protein can interact with a receptor and thus be activated. Accordingly, if the receptor interacts with the ligand (step 2), the protein changes its conformation, thus leading to the bond and the activation of the G protein (step 3), and, in turn, induces in subunit $\alpha$ the exchange between the GDP and the GTP (step 4). The $\alpha$ subunit, linked to the GTP, leaves from the $\beta\gamma$ subunits, which remain linked to each other (step 5), and both the subunit $\alpha$-GTP and the $\beta\gamma$ complex activate a chain of intracellular signals. The $\alpha$ subunit linked to the GTP also exhibits a proper GTPase activity (that can be modulated by other molecules) and breaks the GTP into GDP and phosphate (step 6), thus completing the transduction process. Finally, the $\alpha$-GDP complex gathers with the $\beta\gamma$ complex, and the cycle comes back to the initial inactivated state (step 1) [Lodish et al. (2000)].

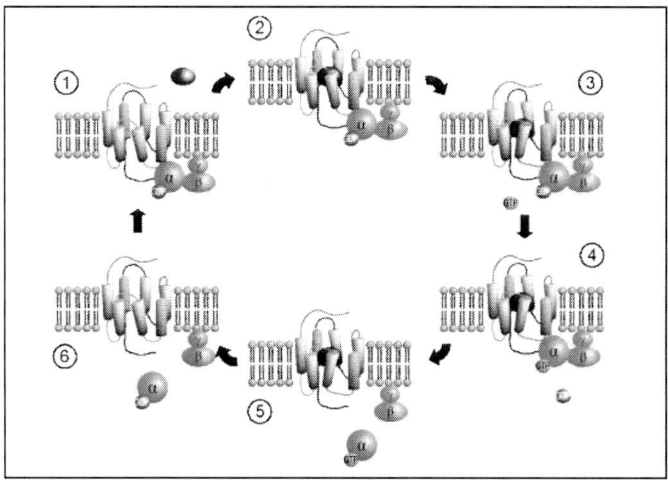


**Figure 2.2** Schematic of the detecting cycle of a GPCR. The arrival and capture of a ligand (1,2) is followed by a conformational change and a consequent release of a G protein (3 to 5) to a final reset of the GPCR (6,1) (Image courtesy: Sven Jähnichen).

### 2.2.1 *GPCR Activation Models*

The classical activation model of GPCRs, and of the consequent signal transmission, consists of a series of events following the ligand capture in the proper bonding sites. The capture determines a conformational change of the receptor, followed by the activation of the G protein and of the effector (channeling or enzymatic). This model is based on the two-state theory [Leff (1995)], which was developed from the following empirical observations.

(i) Some receptors are intrinsically activated, that is, they recall an intracellular signal even in the absence of an agonist.
(ii) There exist complexes, called inverse antagonists, that bind to the receptor and decrease the basic activity of intrinsically activated receptors, thus leading to a response opposite to that of agonists.

On the basis of the above theory, also called lock and key model, any receptor stays in two different states, an inactivated one (R) and an activated one (R*), in equilibrium with each other. Normally, in the absence of the specific ligand, only a small fraction of receptors is intrinsically in the activated state, while all the others are in the inactivated state. The activated state recalls the cell response, acting with a direct cell–cell contact, and even in the absence of an agonist, a given fraction of receptors is in the R* state. Accordingly, there exists a basic cell activity:

$$R \rightleftharpoons R^* \tag{2.1}$$

A given antagonist binds with R or R* with the same affinity, preserving the equilibrium condition. Therefore, it does not call for a response but occupies the receptor inhibiting the interaction with a given agonist [Leff (1995)]. On the other hand, an inverse agonist binds preferentially with an R (inactive), thus inhibiting the basic activity.

Since a long time, these agonists were confused with the antagonists, until the two-state receptor model enabled the mechanism to be explained. The inverse agonists enter the equilibrium between the two different states of the receptor, R and R*, respectively, with a higher affinity for the inactivated state R. As a consequence, the equilibrium shifts to the left, leading to a decrease in the number of activated receptors R* and in the control activity on the effector system. The receptor basic activity is proportional to the receptor number of the active state R*, which justifies the conjecture that the receptor systems with a high number of active states would better evidence the inverse agonism. Finally, the inverse agonists can exhibit different efficacy and become full or partial inverses [Samama et al. (1993)].

Because of the relevant scientific and applied interest on GPCRs, a large quantity of new data have been published in recent years, while a contemporary improvement of knowledge of their three-dimensional structures, as given by X-ray analysis, did not take place [Berman et al. (2000)]. This suggests that the classical activation model is inadequate for explaining the multifaceted functional capacity of GPCRs. In a more updated interpretation, these receptors are part of a more complex mechanism in which their activity is finely modulated by the interaction among different accessory proteins, such as odorant binding proteins [Vidic et al. (2008)]. In particular, the standard on/off switch model of GPCRs activation can be made more realistic by using an energy landscape approach. Accordingly, the space of protein configurations is associated with a continuous energy function whose shape should be described in terms of a double well. The well depths are generally different, with quasi-degenerate minima corresponding to the inactive/active states. In the absence of a specific ligand, the inactive states are more probable and the corresponding well is deeper. The presence of multiple ground states is in agreement with the detected basal activity of the inactive protein. The capture of specific ligands (agonist or inverse agonist) changes the energy landscape allowing some proteins to reach the active state or consolidate in the inactive state [Kobilka (2007); Kobilka and Deupi (2007)]. The dynamic evolution toward a new energy landscape is achieved by means of the breakdown of some noncovalent intermolecular interactions that stabilized the previous receptor configurational state.

The amplitude of the potential energy well is illustrated in Fig. 2.3, which reports the energy landscape of a GPCR. The above scenario reflects the conformation flexibility in a particular energy minimum (see Fig. 2.3A). To better illustrate the different effects that the binding with an agonist or an inverse agonist can produce on the energy landscape, Fig. 2.3B,C) show a proposal of minima shifting in a two-minima model. On one hand, the bond of an agonist should decrease the energy barrier and/or reduce the energy of the most active conformation with respect to the inactive one (Fig. 2.3B). On the other hand, an inverse agonist should produce an increase in the energy barrier and/or a decrease in the conformation of the inactive state that is related to the active conformation (Fig. 2.3C). Of course, more complex modifications of the energy landscape are possible.

In this energetic perspective, the intrinsic difficulty to get X-ray data on the protein structure could be interpreted as due to the simultaneous coexistence of many degenerate configurational states. To this purpose, the nuclear magnetic resonance (NMR) technique should be appropriate to overcome the emerged difficulties.

In conclusion, the proteins belonging to the GPCR family are of crucial importance for the health and the safety of their host organisms. For this reason, since a long time they are object of broad-spectrum investigations, going from the analysis of genetic aspects [Buck and Axel (1991); Crasto et al. (2001); Jones and Reed (1989); Malnic et al. (1999)], like identification and expression, to

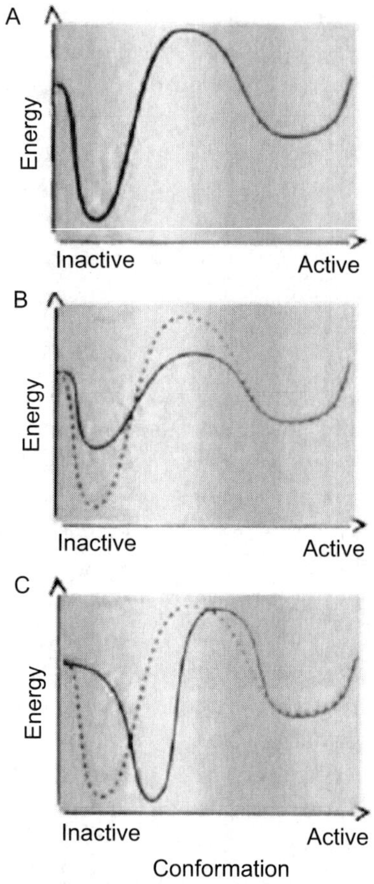


**Figure 2.3** Energy landscape of a GPCR. (A) Conformational state of an unbonded GPCR in the presence of a low basal activity; (B and C) bond of an agonist or of a partial agonist leading to an activity higher than that of the unbound state of panel A (shown as a continuous line). (B) The high basal activity due to a low activation energy barrier is increased because of a reduced energy barrier (compare the new continuous line with the dashed line that reports the curve in A). (C) The high basal activity due to a more active basal conformation is increased because the unbound state has a higher activity toward the G protein (compare the new continuous line with the dashed line that reports the curve in A) [Kobilka (2007); Kobilka and Deupi (2007)].

the planning and realization of different kinds of measurements [Benilova et al. (2008a,b); Gaillard et al. (2004); Hou et al. (2007); Levasseur et al. (2003); Marrakchi et al. (2007); Minic (2005); Minic et al. (2005); Vestergaard et al. (2007); Vidic et al. (2008, 2006)]. Furthermore, the dynamic nature of the GPCRs is probably essential to their physiological functioning, and a better understanding of the molecular plasticity should simplify the pharmaceutic development of new compounds.

### 2.2.2 *Structure and Sensing Action*

Each GPCR, and specifically each OR, is constituted by a given sequence (primary structure) of amino acids (typically a few hundreds). Nevertheless, all these proteins fold up in a quite similar tertiary structure, consisting of seven transmembrane helices (see Fig. 2.1). At present, the tertiary structure of proteins can be experimentally deduced by using both crystallography (X-ray) and NMR techniques [Rhodes (2006)]. These investigation techniques are quite hard, and only for a few (less than 20 [Berman et al. (2000)]) of the recognized ORs, the structure has been measured in the absence of ligand. On the other hand, by using homology modeling, it is possible to produce some plausible structure from a theoretical physicochemical point of view [Fiser and Sali (2003); Launay et al. (2012a,b); Sali and Blundell (1990)]. Specifically, ORs are modeled by using as template the BR whose features have been widely investigated [Palczewski et al. (2000)]. As a matter of fact, BR and ORs exhibit a high sequence homology, and this is sufficient to produce high confidence OR models [Fiser and Sali (2003); Launay et al. (2012a,b); Sali and Blundell (1990)]. Furthermore, the natural abundance and availability of BR, in contrast to other GPCRs typically expressed at a very low level in the cell, places this protein as the pilot receptor for establishing biotechnological platforms for chip-based screening technologies, to characterize the function and interactions of GPCRs [Akimov et al. (2006); Alfinito et al. (2005, 2008, 2009a,c, 2010b); Alfinito and Reggiani (2009b); Hou et al. (2006); Pennetta et al. (2007)].

As previously seen, in a living cell, the detection and transduction processes begin with the conformational change of the GPCRs associated with the capture of the specific ligand. Then, this change activates the G protein, giving rise to a complex sequence of biological mechanisms, which ends with the production of an electric pulse by neurons. It must be noted that different ligands may induce and stabilize distinct conformational states among the various structures that are available and thus promote different G protein activation and receptor desensitization/internalization. Moreover, GPCRs can interact with other proteins through their $C$-terminal domains. They can also give rise to homodimers and heterodimers, with other membrane-bound proteins involved in their functions.

The understanding and in vitro reproducibility of these mechanisms is of great interest for medical and pharmaceutical purposes, in particular for new drug design [Milligan (2004b)]. However, the drug discovery process and the research of novel GPCR-based therapeutics require a large body of data to allow the identification of new receptor ligands.

Therefore, there is an increasing need to develop genetic and chemical high-throughput screening methods for an efficient sorting of compounds with pharmaceutical potential. In this context, investigations pointing to the realization of nanobiosensors based on GPCRs are of strategic importance. Indeed, the production of these devices requires a deep knowledge of the involved GPCRs and, therefore, pushes the research beyond its limits.

In nanobiosensors, the detection process bypasses the complicated sequence of biological events, which follows G protein activation. The goal is rather to achieve a sensitive and reliable monitoring of the conformational change by a direct analysis of the optical and/or the electrical response of the nanodevice itself.

### 2.2.3 *Electrical Characterization*

The electrical properties of living matter and, in particular, of proteins [Weiss (1997)], are object of more and more growing interest [Frauenfelder et al. (1999)]. From the seminal papers on neurons, which established a correspondence with a network of ladder impedances [Weiss (1997)], much work has been performed on both the chemical and physical sides.

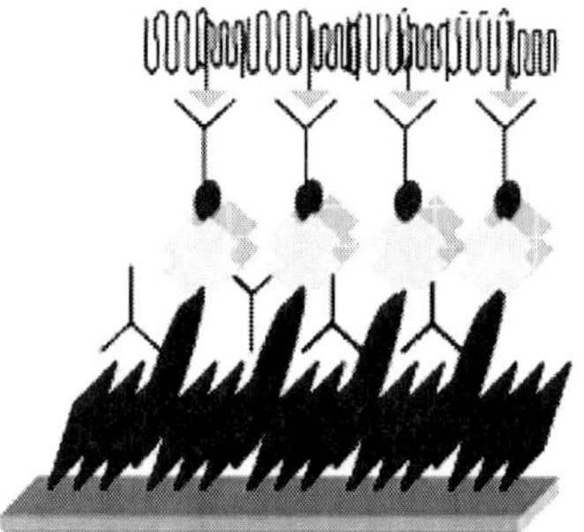


**Figure 2.4** Schematic illustration of the assembling and immobilization process of a typical GPCR (bovine rhodopsin and rat OR I7). Step I: mixed self-assembled monolayers modified gold electrode; Step II: blockage with goat IgG; Step III: binding of neutravidin; Step IV: immobilization of biotinylated antibody Biot-Rho-1D4 [Hou et al. (2006)].

The impedance properties of some GPCRs have been investigated by using both electrochemical impedance spectroscopy (EIS) measurements and also surface plasmon resonance (SPR) technique [Benilova et al. (2008b); Hou et al. (2006, 2007)]. When applied to these proteins, these techniques of investigation mainly point to reveal a change of impedance in the presence/absence of the specific odorant molecules. Preliminary investigations were performed on BR and used for producing a protocol of measurements [Hou et al. (2006)]. Thereafter, the procedure was applied to a couple of ORs, rat OR I7 and human OR 17–40, whose odorant ligands are well documented [Benilova et al. (2008a,b); Duchamp-Viret et al. (1999); Hou et al. (2007); Jacquier (2006); Jacquier et al. (2006); Krautwurst et al. (1998); Levasseur et al. (2003); Marrakchi et al. (2007); Vidic et al. (2008, 2006)].

EIS measurements were carried out on proteins anchored on a self-assembled multilayer (SAM), at different stages of multilayer formation as reported in Fig. 2.4, which illustrates a possible assembling and immobilization process of a typical GPCR [Hou et al. (2006)]. The monitoring of the anchoring process of a GPCR on a substrate was described in terms of Nyquist plots, as shown in Fig. 2.5, which reports the EIS of BR at different levels of assembling on a functionalized gold substrate. Precisely, these plots can be

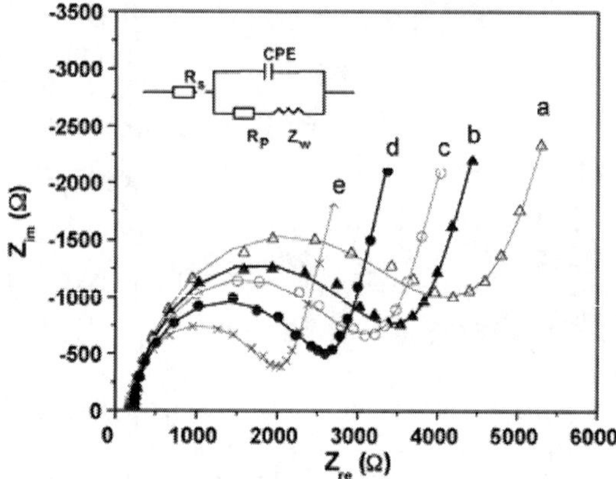


**Figure 2.5** Electrochemical impedance spectroscopy of bovine rhodopsin SAM at different levels of assembling with the corresponding circuital analogue (Randles cell) (on the top). (a) Step I: mixed SAMs modified gold electrode; (b) Step II: blockage with goat IgG; (c) Step III: binding of neutravidin; (d) Step IV: immobilization of biotinylated antibody Biot-Rho-1D4; (e) after injection of 80 ng/ml rhodopsin membrane fraction [Alfinito et al. (2009a, 2010a); Hou et al. (2006)].

interpreted in terms of a simple electrical analogue, the Randles cell, as shown in the inset of Fig. 2.5). This is the circuital description of the macroscopic properties of the electrochemical cell and consists of:

(i) the resistance of the electrolyte solution ($R_S$) related to the highest frequency region (left part of the plot),
(ii) the polarization resistance ($R_P$) and the constant phase elements (CPE) describing the electrode/electrolyte interface features, and
(iii) the Warburg impedance ($Z_W$) related to the lowest frequency region (right part of the plot). Different levels of assembling correspond to different Nyquist plots, but only the semicircular part modifies its shapes. This mean that in the circuit analogue, the series resistance ($R_S$) and the Warburg impedance ($Z_W$) do not change their value significantly as compared with the polarization resistance. Indeed, both $Z_W$ and $R_S$ mainly pertain to the experimental environment. Accordingly, what

really describes the sampling modification is practically only the impedance related to the $R_P$-CPE parallel circuit.

## 2.3 Main Properties of Investigated Proteins

Tables 2.1–2.4 report the specific ligand and the sequential amino acid structure of the proteins that are investigated all along this book.

**Table 2.1** Specific ligand and sequential amino acid structure of proton pump family

| Protein name | Ligand | Amino acids sequence |
|---|---|---|
| Bacteriorhodopsin | Green light | QAQITGRPEWIWLALGTALMGLGTLYFLV<br>KGMGVSDPDAKKFYAITTLVPAIAFTMYL<br>SMLLGYGLTMVPFGGEQNPIYWARYADWL<br>FTTPLLLLDLALLVDADQGTILALVGADG<br>IMIGTGLVGALTKVYSYRFVWWAISTAAM<br>LYILYVLFFGFTSKAESMRPEVASTFKVL<br>RNVTVVLWSAYPVVWLIGSEGAGIVPLNI<br>ETLLFMVLDVSAKVGFGLILLRSRAIFGE<br>AEAPEPSAGDGAAATSD |
| Proteorhodopsin | Green light | MGGGDLDASDYTGVSFWLVTAALLASTVF<br>FFVERDRVSAKWKTSLTVSGLVTGIAFWH<br>YMYMRGVWIETGDSPTVFRYIDWLLTVPL<br>LICEFYLILAAATNVAGSLFKKLLVGSLV<br>MLVFGYMGEAGIMAAWPAFIIGCLAWVYM<br>IYELWAGEGKSACNTASPAVQSAYNTMMY<br>IIIFGWAIYPVGYFTGYLMGDGGSALNLN<br>LIYNLADFVNKILFGLIIWNVAVKESSNA<br>PGGGSHHHHHH |

**Table 2.2** Specific ligand and sequential amino acid structure of GPCRs

| Protein name | Ligand | Amino acid sequence |
|---|---|---|
| Human OR 17–40 | Helional | MQPESGANGTVIAEFILLGLLEAPGLQPV<br>VFVLFLFAYLVTVRGNLSILAAVLVEPKL<br>HTPMYFFLGNLSVLDVGCISVTVPSMLSR<br>LLSRKRAVPCGACLTQLFFFHLFVGVDCF<br>LLTAMAYDRFLAICRPLTYSTRMSQTVQR<br>MLVAASWACAFTNALTHTVAMSTLNFCGP<br>NVINHFYCDLPQLFQLSCSSTQLNELLLF<br>AVGFIMAGTPMALIVISYIHVAAAVLRIR<br>SVEGRKKAFSTCGSHLTVVAIFYGSGIFN<br>YMRLGSTKLSDKDKAVGIFNTVINPMLNP<br>IIYSFRNPDVQSAIWRMLTGRRSLA |
| Bovine rhodopsin | Light | XMNGTEGPNFYVPFSNKTGVVRSPFEAPQ<br>YYLAEPWQFSMLAAYMFLLIMLGFPINFL<br>TLYVTVQHKKLRTPLNYILLNLAVADLFM<br>VFGGFTTTLYTSLHGYFVFGPTGCNLEGF<br>FATLGGEIALWSLVVLAIERYVVVCKPMS<br>NFRFGENHAIMGVAFTWVMALACAAPPLV<br>GWSRYIPEGMQCSCGIDYYTPHEETNNES<br>FVIYMFVVHFIIPLIVIFFCYGQLVFTVK<br>EAAAQQQESATTQKAEKEVTRMVIIMVIA<br>FLICWLPYAGVAFYIFTHQGSDFGPIFMT<br>IPAFFAKTSAVYNPVIYIMMNKQFRNCMV<br>TTLCCGKNPLGDDEASTTVSKTETSQVAPA |
| Rat OR I7 | Odors:<br>octanal,<br>heptanal,<br>nonal | MERRNHSGRVSEFVLLGFPAPAPLRVLLF<br>FLSLLAYVLVLTENMLIIIAIRNHPTLHK<br>PMYFFLANMSFLEIWYVTVTIPKMLAGFI<br>GSKENHGQLISFEACMTQLYFFLGLGCTE<br>CVLLAVMAYDRYVAICHPLHYPVIVSSRL<br>CVQMAAGSWAGGFGISMVKVFLISRLSYC<br>GPNTINHFFCDVSPLLNLSCTDMSTAELT<br>DFVLAIFILLGPLSVTGASYMAITGAVMR<br>IPSAAGRHKAFSTCASHLTVVIIFYAASI<br>FIYARPKALSAFDTNKLVSVLYAVIVPLF<br>NPIIYCLRNQDVKRALRRTLHLAQDQEAN<br>TNKGSKNG |

**Table 2.3** Specific ligand and sequential amino acid structure of GPCRs

| Protein name | Ligand | Amino acid sequence |
|---|---|---|
| Chimpanzee OR 7D4 | Androstenone | MEAENLTELSKFLLLGLSDDPELQPILFG<br>LFLSMYLVTVLGNLLIILAVSSDSHLHTP<br>MYFFLSNLSFVDICFISTTVPKMLVNIQA<br>RSKDISYMGCLTQVYFLMMFAGMDTFLLA<br>VMAYDRFVAICHPLHYTVIMNPCLCGLLV<br>LASWFIIFWFSLVHILLMKRLTFSTVTEI<br>PHFFCEPAQVLKVACSNTLLNNIVLYVAT<br>ALLGVFPVAGILFSYSQIVSSLMRMSSTE<br>GKYKAFSTCGSHLCVVSLFYGTGLGVYLS<br>SAVTHSSQSSSTASVMYAMVTPMLNPFIY<br>SLRNKDVKGALERLLSRADSCP |
| Human OR 7D4 | Androstenone | MEAENLTELSKFLLLGLSDDPELQPVLFG<br>LFLSMYLVTVLGNLLIILAVSSDSHLHTP<br>MYFFLSNLSFVDICFISTTVPKMLVSIQA<br>RSKDISYMGCLTQVYFLMMFAGMDTFLLA<br>VMAYDRFVAICHPLHYTVIMNPCLCGLLV<br>LASWFIIFWFSLVHILLMKRLTFSTGTEI<br>PHFFCEPAQVLKVACSNTLLNNIVLYVAT<br>ALLGVFPVAGILFSYSQIVSSLMGMSSTK<br>GKYKAFSTCGSHLCVVSLFYGTGLGVYLS<br>SAVTHSSQSSSTASVMYAMVTPMLNPFIY<br>SLRNKDVKGALERLLSRADSCP |
| OR 2AG1 | Amyl butyrate | MELWNFTLGSGFILVGILNDSGSPELLCA<br>TITILYLLALISNGLLLLAITMEARLHMP<br>MYLLLGQLSLMDLLFTSVVTPKALADFLR<br>RENTISFGGCALQMFLALTMGGAEDLLLA<br>FMAYDRYVAICHPLTYMTLMSSRACWLMV<br>ATSWILASLSALIYTVYTMHYPFCRAQEI<br>RHLLCEIPHLLKVACADTSRYELMVYVMG<br>VTFLIPSLAAILASYTQILLTVLHMPSNE<br>GRKKALVTCSSHLTVVGMFYGAATFMYVL<br>PSSFHSTRQDNIISVFYTIVTPALNPLIY<br>SLRNKEVMRALRRVLGKYMLPAHSTL |
| Canine OR 5269 | Hexanal | MNRSATHIHVVTEFVLLGFPGCWEIQIFL<br>FSFFLVVYVLTLLGNGTIICAVRWEPRLH<br>TPMYFLLGNFAFLEIWYASSTVPNVLANI<br>LSKTKAISFSGCFLQFYFFFSLGTTECLF<br>LAVMAYDRYLAICHPLHYPTVMTGKLCRT<br>LVSLCWLTEFLGYPIPIFLISQLPFCGSN<br>IIDHFLCDMDPLMALSCAPAPITELIFYT<br>QSSLVLFFTIMYILRSYILLLRAVFLVPS<br>AAGRRKAFSTCGSHLAVVSLFYGTVMVMY<br>VSPTYGIPTLMQKILTLVYSIMTPLFNPL<br>IYSLRNKDMKLALRNILFRMRISQNS |

**Table 2.4** Specific ligand and sequential amino acid structure of other proteins studied

| Protein name | Ligand | Amino acid sequence |
|---|---|---|
| Azurin | Cu | AECSVDIQGNDQMQFNTNAITVDKSCKQF<br>TVNLSHPGNLPKNVMGHNWVLSTAADMQG<br>VVTDGMASGLDKDYLKPDDSRVIAHTKLI<br>GSGEKDSVTFDVSKLKEGEQYMFFCTFPG<br>HSALMKGTLTLK |
| AChE | Acetylcholine<br>several reversible<br>and quasi-irreversible<br>inhibitors<br>[Bourne (2003); Colovic et al. (2013)] | GREDAELLVTVRGGRLRGIRLKTPGGPVS<br>AFLGIPFAEPPMGPRRFLPPEPKQPWSGV<br>VDATTFQSVCYQYVDTLYPGFEGTEMWNP<br>NRELSEDCLYLNVWTPYPRPTSPTPVLVW<br>IYGGGFYSGASSLDVYDGRFLVQAERTVL<br>VSMNYRVGAFGFLALPGSREAPGNVGLLD<br>QRLALQWVQENVAAFGGDPTSVTLFGESA<br>GAASVGMHLLSPPSRGLFHRAVLQSGAPN<br>GPWATVGMGEARRRATQLAHLVGCPPGGT<br>GGNDTELVACLRTRPAQVLVNHEWHVLPQ<br>ESVFRFSFVPVVDGDFLSDTPEALINAGD<br>FHGLQVLVGVVKDEGSYFLVYGAPGFSKD<br>NESLISRAEFLAGVRVGVPQVSDLAAEAV<br>VLHYTDWLHPEDPARLREALSDVVGDHNV<br>VCPVAQLAGRLAAQGARVYAYVFEHRAST<br>LSWPLWMGVPHGYEIEFIFGIPLDPSRNY<br>TAEEKIFAQRLMRYWANFARTGDPNEPRD<br>PKAPQWPPYTAGAQQYVSLDLRPLEVRRG<br>LRAQACAFWNRFLPKLLSAT |

# Chapter 3

# Electrical Properties: Experiments

## 3.1 General

Different techniques have been reported in the literature to investigate the electrical properties of proteins. In this field, optical techniques play an important role, which, in most cases, make use of visible fluorescent proteins (VFPs) [Lidke et al. (2004)] and of confocal microscopy analysis, like in fluorescence resonance energy transfer (FRET) and fluorescence lifetime imaging (FLIM) techniques [Pompa et al. (2004); Wallrabe (2005); Wallrabe et al. (2002)]. Other important tools for the investigation of this kind of systems are offered by the surface plasmon resonance (SPR) technique [Bieri et al. (1999); Neumann et al. (2002); Vidic et al. (2008, 2006)] and by atomic force microscopy (AFM) [Andolfi (2006); Bizzarri and Cannistraro (2012); Kivioja et al. (2009); Raccosta et al. (2013); Wu et al. (2009, 2012); Zhao et al. (2004)]. In particular, the AFM technique represents a very powerful tool for structural biology studies since it gives access to the molecular architecture [Biasco (2004); Fotiadis et al. (2002)] and it can be used to follow and characterize receptor immobilization on selected biostructured surfaces.

*Proteotronics: Development of Protein-Based Electronics*
Eleonora Alfinito, Jeremy Pousset, and Lino Reggiani

ISBN 978-981-4613-63-7 (Hardcover), 978-981-4613-64-4 (eBook)
www.panstanford.com

Among the experimental methods used to investigate the electrical properties of proteins and their conformational change during a sensing action, the following five are considered more relevant and promising in the near future: (i) electrochemical impedance spectroscopy (EIS), (ii) carbon nanotube (CNT) transistor, (iii) metal–protein–metal (MPM) system in thin film structures or (iv) in nanolayer film structures, and (v) AFM with nanometric resolution.

Electrochemical impedance spectroscopy has proven to be one of the most effective techniques for the characterization of biomaterials deposited on functionalized electrodes and of biocatalytic transformations at electrode surfaces and, specifically, for the detection of biosensing events at electrodes [Guan et al. (2004); Hou et al. (2006, 2007); Katz and Willner (2003); Pei et al. (2001); Rickert et al. (1996)]. When compared to other electrochemical techniques, one of the great advantages of EIS is the small amplitude of the perturbation signal from the thermal equilibrium steady state, which allows to describe the response as quasi-linear [Barsoukov and Macdonald (2005)]. Therefore, EIS has been extensively used to characterize the process of fabrication of biosensors and to monitor biomolecular recognition [Beyer et al. (1994); Manickam et al. (2012)]. Very recently, EIS measurements have demonstrated the sensitivity and selectivity of self-assembled multilayer (SAM) systems for the specific grafting of bovine rhodopsin membrane fraction and other G protein–coupled receptors (GPCRs) [Bourigua et al. (2013); Hou et al. (2006, 2007); Jaffrezic-Renault (2013)].

The use of carbon nanotubes to immobilize the protein on a field-effect transistor was proven to be an ideal strategy for the easy and fast detection of odorants identified by the change in the electrical characteristics of the appropriate OR [Kim et al. (2009); Park et al. (2012)]. This important feature was exploited to build a portable electronic device operating as a bioelectronic nose [Park et al. (2012)].

The use of MPM structures based on micrometer thin films (typically of pR) [Melikyan et al. (2011)] or of a nanometer layer-width of the protein under study (typically bR) [Jin et al. (2006)] proved the possibility to detect the photoconductive property of both bR and pR. The nanostructures offer the advantage of applying very high electric fields, of the order of MV/cm, thus exploring the nonlinear behavior of the current–voltage characteristics.

The use of AFM technique satisfies the possibility to make local nano-measurements and to apply ultra-high electric fields (over 10 MV/cm) and thus to investigate the microscopic mechanisms of charge transfer in proteins, typically dominated by tunneling processes [Casuso et al. (2007a,b)].

The following sections provide brief illustrations of the basic features of each of the techniques listed above.

## 3.2 Electrochemical Impedance Spectroscopy

### 3.2.1 *Model Lipid Bilayer*

The major constraint to carry out a reliable EIS measurement is to use samples in which proteins preserve structure and function. With reference to transmembrane proteins, and in particular to GPCRs, they are extremely dependent on the membrane environment to maintain their proper functions. This represents a challenging problem in building artificial protein scaffolds, which has been faced with several different strategies, including the use of micelles, bicelles, lipid vesicles, nanodiscs, lipidic cubic phases, and planar lipid membranes (for a recent review, see [Serebryany et al. (2012)]).

The first step toward the preparation of reconstituted samples is to prepare and purify a sufficient quantity of the protein under examination. Membrane receptors are usually expressed in heterologous cells, solubilized and purified in an appropriate detergent [Bieri et al. (1999); Rebois et al. (2002)]. In a few cases, proteins can be purified directly from natural sources. This happens for BR that was initially purified from preparations of bovine retinas [Okada et al. (1998)]. Although many such successful procedures are reported in the literature, the method requires substantial effort for the purification of GPCRs prior to analysis. The choice of expression systems, such as yeast and insect cells, depends on the GPCR under study [McCusker et al. (2007)].

After the expression and the subsequent purification, the reconstitution of proteins in a lipid membrane can proceed. The most basic strategy for in vitro studies of purified GPCRs is the use of detergent or *detergent-mixed micelles* [Serebryany et al. (2012)]. Micelles are assemblies of amphipathic molecules with their hydrophilic heads exposed to solvent and their hydrophobic tails in the center. Micelles can solubilize membrane proteins by partially encapsulating them and shielding their hydrophobic surfaces from solvent. Although this method produces some good results in the stabilization of proteins, it has some relevant drawbacks especially for BR. First, the micelle is a highly disordered environment compared to the native membrane. Second, the concentration of free detergent in solution is high and then it may interfere with normal ligands and G protein binding. In general, it is difficult to use a system of this kind for studying the interaction between a GPCR and its soluble protein-interacting partners, since the fold and function of the soluble protein may not be preserved in a detergent [Serebryany et al. (2012)].

In detergent-solubilized form, GPCRs can be reconstituted in *unilamellar lipid vesicles*. These can be of small size (about 100 nm) and are called nanosomes or small unilamellar vesicles (SUVs), or of quite large size (up to 100 μm), and are called giant unilamellar vesicles (GUVs). The vesicles are prepared by solubilizing the protein samples and phospholipids with cholate and subsequently removing the detergent by gel filtration and collecting the resulting vesicles. Using a similar procedure, the entire primary pathway of cholinergic signaling in muscarin receptor was reconstituted [Serebryany et al. (2012)]. Recent application of this technique has allowed the expression of BR and some ORs [Hou et al. (2006); Minic et al. (2005)]. Finally, GUV provides a key model membrane system to study lipid–lipid and lipid–protein interaction. As a matter of fact, due to their large size, these samples can be investigated with several techniques such as fluorescent or confocal microscopy [Wesolowska et al. (2009)]. The above procedure has been used with bacteriorhodopsin, which has a close structural homology with GPCRs. So far, no GPCRs have been yet reconstituted using this method, which still remains a very promising one.

*Bicelles* are fragments of a lipid bilayer whose perimeter is stabilized by short-chain lipids or detergent molecules [Serebryany et al. (2012)]. The bicelle is a versatile system for GPCR reconstitution, thanks to the wide range of lipid compositions it can accommodate [Serebryany et al. (2012)]. With respect to mixed micelles, it allows reduced free detergent concentrations, thus reducing structural perturbation of soluble macromolecules. Otherwise, for proteins such as GPCRs, the preparation can induce a quite high percentage of empty bicelles. The bicelle system has been used to obtain the first detailed characterization of the kinetics of retinal binding to the opsin protein to form bovine rhodopsin.

*Nanodiscs* are the most novel paradigm for GPCR reconstitution. They resemble bicelles but provide a more stable and well-defined membrane environment. Nanodiscs consist of a lipid bilayer center, which can incorporate a transmembrane protein, and two molecules of membrane scaffold protein (MSP), a helical repeat protein with hydrophobic and hydrophilic faces. MSP wraps around the hydrophobic edge of the lipid disc and stabilizes it in an aqueous environment [Serebryany et al. (2012)]. The $\beta_2$-adrenergic receptor [Leitz et al. (2006)] and CCR5 chemokine receptor [Knepp (2011)] have been successfully incorporated into nanodiscs, and it has been demonstrated that this sample can activate G proteins. Purified BR has also been reconstituted using this system, demonstrating both light-induced activation of the receptor and activation of the G protein transducing or arresting as a result.

Since a long time, *planar lipid membranes* are known as black lipid membrane (BLM), referring to membranes created in a thin layer of a hydrophobic material such as Teflon. These membranes are affected by two main problems: residual solvent and limited lifetime. Another way to produce planar lipid membrane is laying them on a solid support, a solid lipid membrane (SLM). In this way, only the upper face of the bilayer is exposed to free solution. SLMs remain largely intact even when subject to high flow rates or vibrations and, unlike BLMs, the presence of holes will not destroy the entire bilayer. Because of this stability, long-lasting (weeks or months) experiments can be performed, while BLM experiments are usually limited to hours. A remarkable advantage of SLM structures is that, since they have a flat and hard surface, several different characterization tools, which would be impossible or would offer lower resolution if performed on a freely floating sample, can be adopted. AFM technique has been used to image lipid phase separation, formation of transmembrane nanopores followed by single-protein molecule adsorption, and protein assembly with sub-nanometer accuracy without the need for a labeling dye [Tokumasu et al. (2002)]. Many modern fluorescence microscopy techniques also require a rigidly supported planar surface. Evanescent field methods, such as total internal reflection fluorescence microscopy (TIRFM) and SPR, can offer extremely sensitive measurement of analyte binding and bilayer optical properties but can only function when the sample is supported on specialized optically functional materials. Another class of methods applicable only to supported bilayers is that based on optical interference, such as fluorescence interference contrast microscopy (FLICM) and reflection interference contrast microscopy (RICM).

*Lipid cubic phase* (LCP) is a sponge-like membrane system consisting of lipid, water, and protein in appropriate combinations. It forms a structured, transparent, and complex three-dimensional lipidic array, which is pervaded by an intercommunicating aqueous channel system [Landau and Rosenbusch (1996)]. Ordinarily, cubic phases are less ordered than a solid but more ordered than a liquid. A lipidic cubic phase containing cholesterol has been used to reconstitute and crystallize several rceptors such as $\beta_2$-adrenergic, adenosine A2A, dopamine $D_3$, and CXCR4 chemokine [Serebryany et al. (2012)]. This method allows a high concentration of proteins and, like vesicles, the free diffusion of reconstituted proteins. Among proteins, native-like interactions are also detected, while the ligand-binding and G protein–binding surfaces of a GPCR are more difficult in a cubic phase than in nanodiscs or planar lipids membranes.

### 3.2.2 *Immobilization of GPCRs*

In what concerns the immobilization of GPCRs, the SAM technique appears to be the most suitable and effective for the construction of well-ordered and ultra-thin organic films, since it allows the required control to be carried out at a molecular level. In short, SAMs of organic molecules are molecular assemblies spontaneously formed on metallic surfaces (often Au) by adsorption and are organized into highly ordered domains [Nuzzo and Allara (1983)]. In general, they consist of a head group (often thiol, but also silane or phosphonate), a tail, and a functional group on which the substrate can attach. Successive investigations, mainly devoted to the development of biosensors [Ferretti (2000); Hou et al. (2006, 2007); Jaffrezic-Renault (2013); Samanta and Sarkar (2011); Vericat et al. (2010)], successfully adopted an evolution of SAM technique, the so-called self-assembled multilayer technique [Love (2005); Ulman (1996)]. Due to the simplicity in production, the adaptability and the possibility of controlling the orientation of biomolecules on the surface, this technique stands out as the most effective in the production of artificial biomolecular surfaces with detection aims. In particular, as already sketched in Fig. 2.4, the protein supporting structure was obtained as follows. A gold substrate adsorbs MHDA thiol and biotinyl-PE. Then the antibody goat IgG stabilizes the film, also saturating all the nonspecific adsorption sites, and blocks a layer of neutravidin and biotinylated polyclonal antibody specific for the protein to immobilize [Hou et al. (2006, 2007); Minic et al. (2005)]. In particular, it was found that the avidin-biotin system works very well as a bridge to anchor a bioreceptor, since the biotinylation of a biomolecule does not affect its biological activity. Furthermore, the noncovalent complex between avidin and biotin is characterized by a very high affinity constant of $10^{15}$ $mol^{-1}$ l. Once formed, the bond is stable even if the pH of the solution is changed and it can easily resist multiple washings [Storri et al. (1998)]. Neutravidin was then used to anchor biotin-labeled antibodies of the specific receptor. In biosensor research, such a multilayer system using biotin/avidin pairs acting as binding agents is of relevance to immobilize a membrane receptor. Immobilization of the prototype GPCR, BR, in its native membranes, was thus achieved on functionalized surfaces [Hou et al. (2006); Minic et al. (2005)], which represents the prerequisite for elaborating a GPCR-based biosensor. By using the same technique and specific biotinylated antibodies, the rat OR I7 [Hou et al. (2007)] and the human OR 17–40 [Benilova et al. (2008a,b)] were also successfully immobilized and analyzed.

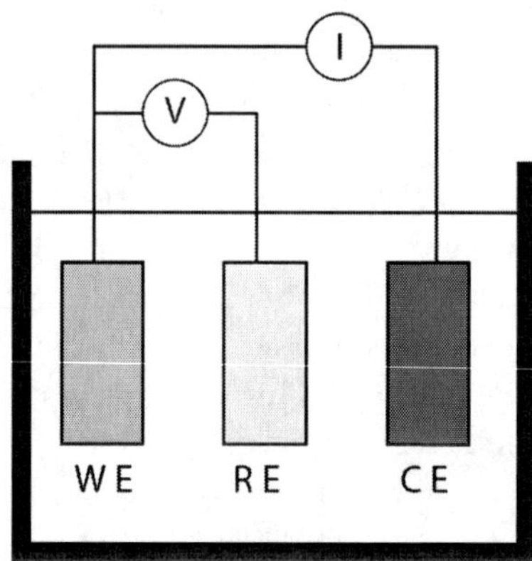


**Figure 3.1** Schematic of a three-electrode electrochemical cell. WE, RE, and CE refer, respectively, to working, reference, and control electrodes.

## 3.2.3 *Experimental Results*

Figure 3.1 shows a schematic of a typical electrochemical cell used for EIS experiments. The target with the protein anchored to the given substrate is attached to the working electrode with a typical active surface of a fraction of $cm^2$. All the solutions inside the cell are prepared in a phosphate buffer solution (PBS) (1.8 mM $KH_2PO_4$, 0.1 mM $Na_2HPO_4$, 140 mM NaCl, and 2.7 mM KCl, pH 7.0). After each step, the electrodes are thoroughly washed with PBS to remove non-specifically adsorbed biomolecules. A given AC current is imposed between the working and the counter-electrode, and the potential drop is measured between the working electrode and a reference electrode. Frequency values of the applied fields are in the typical range 1 mHz to 100 KHz. Typical EIS spectra are reported as Bode plot and/or as Nyquist plot. The former provides complete representation of the small-signal impedance in two graphs that report the modulus and the phase, respectively, as a function of the frequency. Figure 3.2 reports an example of the Bode plot for the case of rat OR I7 during the characterization of a sample at given concentrations of OR membranes. The experiments are best fitted with a standard software (Z-plot/Z-view software, Scribner Associates, Inc.) using an equivalent circuit (the Randles cell) appropriate to this case. The latter, also called Cole–Cole plot,

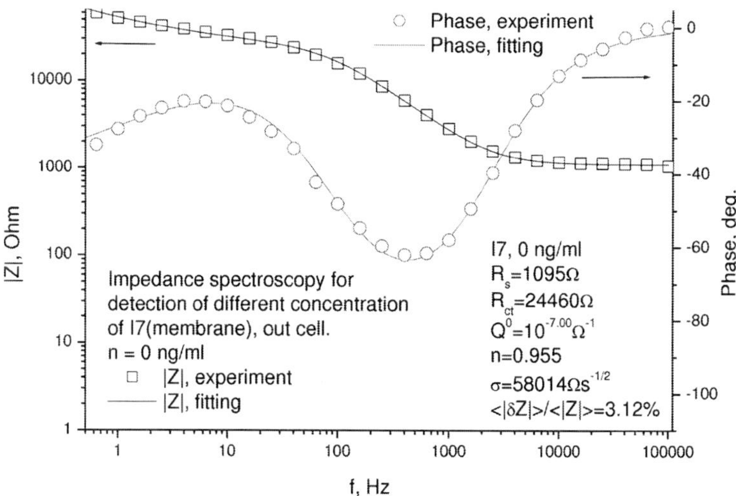


**Figure 3.2** Bode plot of an EIS experiment carried out using a rat OR I7.

reports minus the imaginary part of the impedance as function of the real part of the impedance with high (low) frequency regions being associated with the low (high) values of the real part of the impedance. This graph combines in a single plot the two graphs that compose the Bode plot. An example of Nyquist plots is reported in Fig. 3.3 for the case of rat OR I7 during the characterization of a sample at increasing concentrations of octanal, a specific odorant of rat OR I7. Even in this case, the experiments are best fitted by using the procedure previously described.

## 3.3 Carbon Nanotube Field-Effect Transistor

In the following, the experiment of Park et al. (2012) is described, a prototype toward the realization of olfactory–nanovesicle-fused carbon nanotube–transistor biosensors (OCBs). In particular, the specific device mimics some features of a dog nose and is able to detect hexanal, an indicator of food oxidation, at very low concentrations (down to 1 fM).

A schematic diagram depicting the fabrication method of an OCB is reported in Fig. 3.4A [Park et al. (2012)]. By using a standard

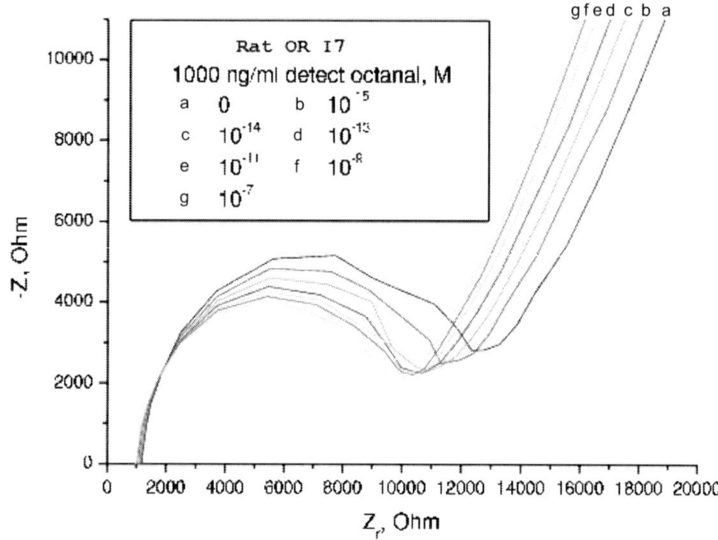


**Figure 3.3** Set of Nyquist plots carried out for the rat OR I7 in the presence of increasing doses of the specific odorant octanal.

procedure [Lee et al. (2006); Park et al. (2012)], OCB can be obtained by selectively adsorbing CNTs onto a bare $SiO_2$ surface region. The CNT patterns appear random. Then this structure is equipped with Ti/Au electrodes, fabricated by using the photolithography process. Both metallic and semiconducting CNTs are used. In this case, the transistors based on such CNTs do not turn off completely and are not suitable for integrated circuit applications. However, they exhibit a transconductance large enough for sensor applications. Finally, in the channel region of the CNT transistor, nanovesicles containing the canine olfactory receptor cf OR 5369, the related G proteins and enzymes, and finally calcium channels were immobilized. Since the shape of nanovesicles could be sustained only in a solution, OCBs should not be stored for over a day [Park et al. (2012)].

For the sensing experiments, the OCB is connected by source and drain electrodes; then it is covered with an appropriate liquid solution [Lee et al. (2011)]. An AC measurement based on a lock-in amplifier is utilized to monitor the source and drain currents as sensor signal after applying the odorant (hexanal, specific for cf OR

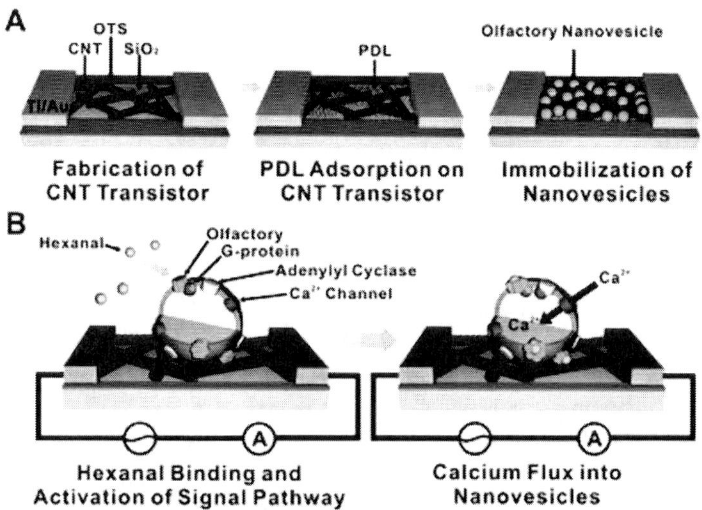


**Figure 3.4** OCB for the detection of hexanal. (A) Schematic diagram showing a method to prepare an OCB. A CNT was coated with poly-D-lysine for the stable adsorption of nanovesicles. Then olfactory nanovesicles were immobilized on a CNT channel region in the transistor. (B) Schematic diagram showing the sensing mechanisms for the detection of hexanal with an OCB. The binding of hexagonal to ORs results in a $Ca^{2+}$ influx into the nanovesicles through $Ca^{2+}$ channels. Here the accumulated $Ca^{2+}$ ions inside the nanovesicles create a positive gate potential in the vicinity of underlying CNTs, and the increased potential results in the decrease of conductance in the CNT channel [Park et al. (2012)]. [Reproduced with permission of The Royal Society of Chemistry].

5369) solution with different concentrations. In this measurement scheme, an alternating voltage (30 mV, 31 Hz) is applied to the source and drain electrodes of an OCB by a function generator. The current through the electrodes is amplified by a low-noise current preamplifier, and a lock-in amplifier is employed to measure the amplitude of the output signal of the preamplifier. This AC operation using a lock-in amplifier enables high-precision measurement of the sensor signals to be carried out without being disturbed by environmental noises with different frequencies. Previously, a lock-in amplifier was utilized to measure the conductance of a field-effect transistor (FET) sensor and detect biological and chemical species [Cui et al. (2001)]. In this case, low operation frequencies such as

31 Hz are welcome in order to allow the CNT-FET sensors to exhibit a behavior close to that of DC mode operation.

Figure 3.4B depicts the mechanism for the detection of hexanal using the OCB device. The binding of hexanal on ORs successively activated G proteins, adenylyl cyclases enzyme, and $Ca^{2+}$ channels following the cAMP pathway in the nanovesicle [Firestein (2001)]. The influx of $Ca^{2+}$ ions through the channels increases the potential of the nanovesicle in the vicinity of CNTs. Since CNT channels exhibit a p-type behavior under ambient conditions, the increased potential of the nearby nanovesicle results in the decrease of the CNT channel conductance. In such a way, the complete pathway of signal transmission is converted in an electrical signal and, as a final result, the detection of odorant is performed with high sensitivity and selectivity. The image of fixed nanovesicles on the CNT channel region is taken by a scanning electron microscope (SEM) with a 30 kV accelerating voltage and a $\times 13000$ magnification.

Figure 3.5 reports real-time conductance measurements obtained from an OCB after the introduction of various odorants. The addition of 1 nM pentanal, heptanal, octanal, and hexanal does not affect the conductance of the OCB, while the addition of 1 pM of hexanal causes a sharp decrease in the conductance. Note that the difference between those molecule odors is very minor. For example, the difference between hexanal and heptanal is just a single carbon atom in their alkane chains. This result shows that OCB devices can detect hexanal with a high selectivity.

## 3.4 Metal–Protein–Metal Structure: Thin Film Technique

Melikyan et al. illustrated the use of MPM thin film technique [Melikyan et al. (2011)]. The preparation of the sample, which contains pR as the active protein sensitive to visible light, proceeds as follows. First, proteins are expressed in *Escherichia coli* by using recombinant DNA technology [Choi et al. (2007)]. Second, they are purified, dialyzed, and concentrated. Third, pR is diluted in a buffer solution (50 mM Tris, 150 mM NaCl, and 0.02% dodecyl-$\beta$-D-maltoside with pH 9) with increasing amounts: 0, 0.2, 0.5, 1, 2, 5, and 10 OD (optical density).

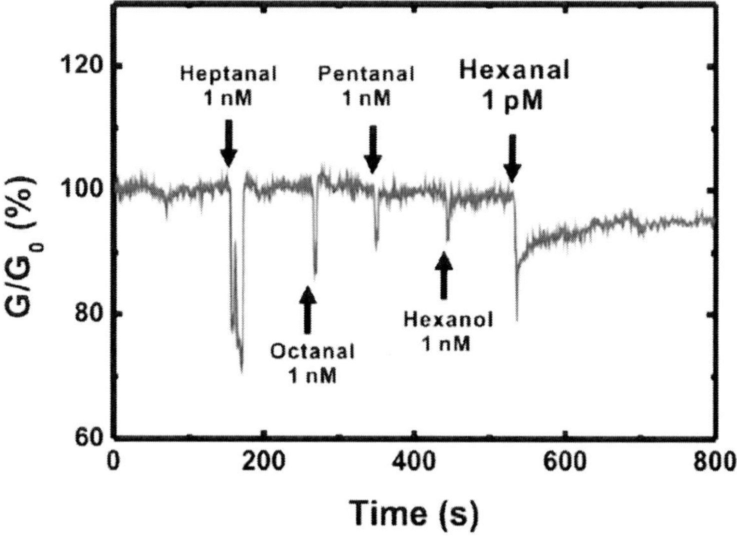


**Figure 3.5** Real-time conductance measurements obtained from an OCB after the injection of different odorants. The addition of 1 nM heptanal, octanal, pentanal, and hexanal solutions had no effect on the conductance of the OCB, while the addition of 1 pM of hexanal solution caused a sharp decrease in the conductance of the OCB normalized to its unperturbed value $G_0$. [Reproduced with permission of The Royal Society of Chemistry].

The OD-532 to mol conversion is given by the Beer–Lambert law at 532 nm wavelength: $A = \alpha lC$, where $A$ is the absorbance in the optical density [no units, $A = \log(P_0/P)$], $\alpha$ is the molar absorptivity or extinction coefficient (l/molcm), $l$ is the path length of the sample (cm), and $C$ is the concentration of the compound in solution (mol/l). For pR, $\alpha = 42.884$ l/molcm, $l = 1$ cm, and $C = 2.33 \times 10^{-5}$ mol/l. Note that 1 OD corresponds to 0.629 mg/ml concentration of pR.

The experimental setup of the MPM technique is shown in Fig. 3.6. The pR sample consists of a cell for containing the buffer plus the protein solution and was obtained by evaporating onto a glass (transparent) plate, two 200 nm thick gold electrodes. The channel was 50 μm width, 3 mm length. The top of the glass/Au surface was covered by pR solution with different protein concentrations and afterward dried at 40°C for 30 min. All samples

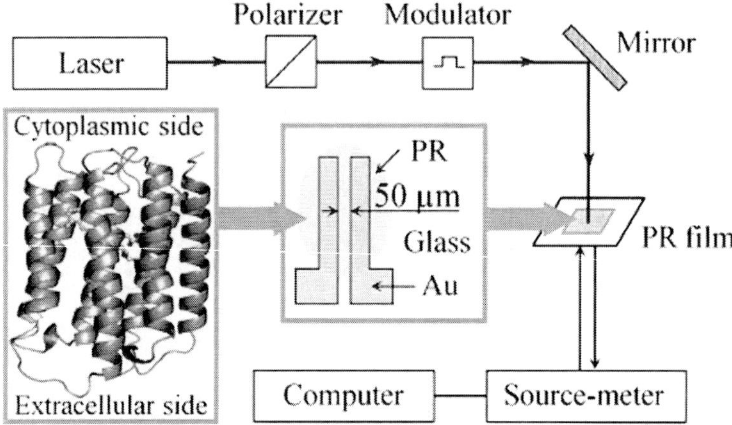


**Figure 3.6** Thin film MPM experimental setup, pR sample configuration, and three-dimensional structure of pR drawn as a ribbon diagram with the retinal chromophore [Melikyan et al. (2011)].

are prepared so that the resulting dry films had similar thicknesses around 2 μm. The pR conducting properties are investigated under different conditions of illuminations (from dark to blue light, intensity up to 150 mW/cm$^2$).

The experiment is performed by using linear-polarized coherent light emitted by a laser source. Current–voltage measurements are carried out by a source-meter, at room temperature (25°C) and at the ambient relative humidity of about 40%.

The results of the photoresponse experiment of the pR films upon illumination are reported in Fig. 3.7, where the typical time response under laser pumping modulation upon illumination is shown in the inset of the same figure. As a remarkable result, the pR film exhibits a pronounced photocurrent (about 10 nA, with an applied bias of 120 V) under different illumination conditions. In particular, photocurrent is maximized by the 532 nm wavelength light and increases with the pR concentration in the sample, signaling that the main conductance is due to the presence of the protein. As apparent from Fig. 3.7, the I–V characteristic is almost linear for high (>2 OD) pR concentrations, which means that the conductivity is constant at fixed light intensity and wavelength. The photocurrent exposure time (ET) also exhibits approximately

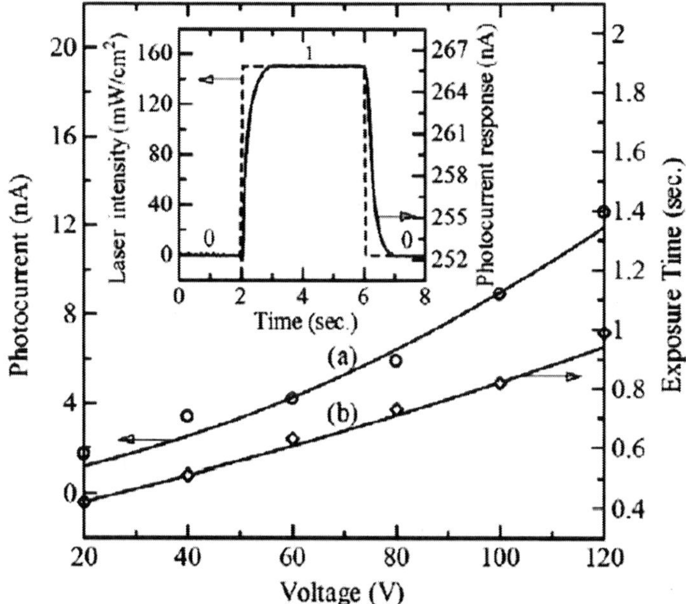


**Figure 3.7** (a) The photocurrent amplitude (left axis) and (b) exposure time (right axis) versus applied bias for 10 OD pR under illumination with the intensity of 150 mW/cm$^2$ at 635 nm. The inset shows the pumping laser pulse modulation (dashed line) and pR photocurrent (solid line) time response with 120 V applied bias. The position 0 and 1 are off and on states of the photogeneration, respectively [Melikyan et al. (2011)].

a linear behavior. Current ET or kinetics is the effective length of time a current takes to rise to the maximum level under illumination. Probably, the photocycle time (limited by the decay of two intermediates, normally faster M-like and slower O-like) increases up to a few hundred milliseconds in dry pR film, similar to what is known for bR and prolongs the ET.

## 3.5 Metal–Protein–Metal Structure: Nanolayer Technique

To illustrate this technique, Ref. [Jin et al. (2007, 2006)] is used. Here it has been noticed that although bR is naturally found in a two-dimensional membrane (5 nm thick), which could be the first choice for measurements, the small size of PM natural patches (few μm) does not allow production of samples with an efficient coverage. This, in turn, makes difficult to obtain MPM devices ensuring trustworthy and reproducible measurements [Jin et al. (2007)]. Therefore, the preparation of the sample containing bR as active protein sensitive to visible light is obtained following this strategy: Purple membrane fragments containing wild-type bR were prepared by using standard techniques and reconstituted in vesicles. A substrate of Al with a layer of Al oxide ($AlO_x$) on its surface was immersed in a vesicular suspension in such a way that the vesicles, adsorbed by the substrate fused and opened, formed a monolayer. By using the lift-off float-on (LOFO) technique [Jin et al. (2006)], gold dot contacts (60 nm thick, $2.0 \times 10^{-3}$ cm$^{-2}$ wide) were placed on the monolayer. The advantage of LOFO with respect to standard techniques is the possibility of contacts to float on the monolayer, so bridging small defects such as pinholes and cracks and allowing more stable electrical measurements [Jin et al. (2007, 2006)]. The resulting structures were able to sustain multiple and reproducible measurements of electron transport, in dark and light conditions. The environmental conditions (humidity and room temperature) were not modified. I–V measurements were carried out on planar junction structures in a class-10000 clean room at 293 K and 40% relative humidity, with the sample sandwiched between the Al/$AlO_x$ substrate and the Au contacts. The circuit was completed by gently placing a tungsten electrode on the gold electrode, as schematically reported in Fig. 3.8 [Jin et al. (2006)].

Figure 3.9 reports typical I–V characteristics of a resulting metal/bR monolayer-in-lipid bilayer/metal junction over a $\pm 1$ V bias range. The I–V measurements were performed in dark and light, and in both cases, the current was found to grow monotonically at increasing voltages, with a marked superlinear behavior. The maximal current value of 0.75 nA was recorded after the application of 1 V bias in dark. After steady-state illumination with green light ($\lambda > 550$ nm, 20 mW/cm$^2$), the current increased from 0.75 to 1.7 nA at an applied bias of 1 V. Once the green light was switched off, the current decayed over 2–3 min to its original dark value.

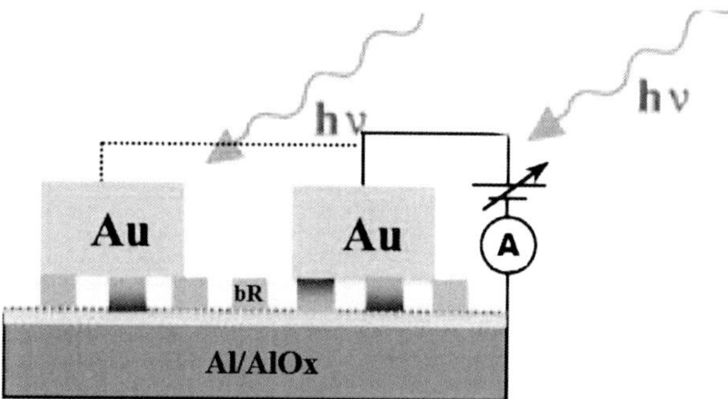


**Figure 3.8** Schematic of the MPM nanostructure for I–V measurements in the presence of light [adapted from Jin et al. (2006)].

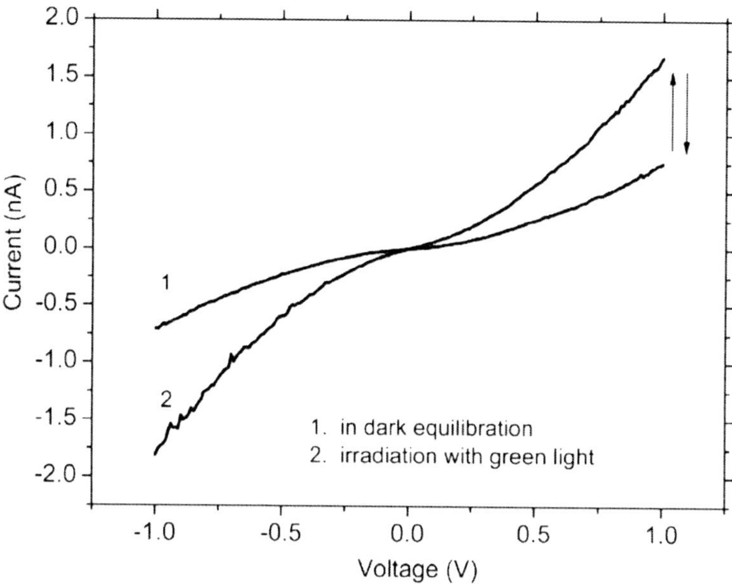


**Figure 3.9** I–V characteristics of bacteriorhodopsin nanolayers in dark and green light [adapted from Jin et al. (2006)].

## 3.6 Atomic Force Microscopy Technique

To illustrate this technique, Ref. [Casuso et al. (2007a,b)] is used. Here the preparation of the sample, which contains bR as active protein sensible to visible light, proceeds as follows. Purple membranes containing wild-type bR are isolated from *Halobacterium salinarum* to a final concentration of $10^{-7}$ M in Milli-Q water. PM patches are deposited on a gold film evaporated on a mica substrate and dried under $N_2$(g), in such a way as to preserve the PM functionality [Váró and Keszthelyi (1983)]. Finally, the samples are inserted in an AFM chamber in $N_2$(g) atmosphere, where the measurements are performed. Notice that the small-sized PM patches are sufficiently large for AFM investigations (tip nominal size 100–200 nm).

Conductive-AFM (C-AFM) technique combines topographical AFM and electrical DC measurements and proceeds as follows. The C-AFM setup is composed of a commercial AFM with the AFM probe connected to a full-customized current-to-voltage amplifier. While the AFM probe is maintained at virtual ground, the substrate is connected to a DC bias applied by the AFM control unit. The current flow coming from the AFM probe is measured by the amplifier and recorded simultaneously to the rest of AFM signals.

Figure 3.10 reports a schematic of the AFM setup with the detection chain. The amplifier is fully customized for AFM measurements

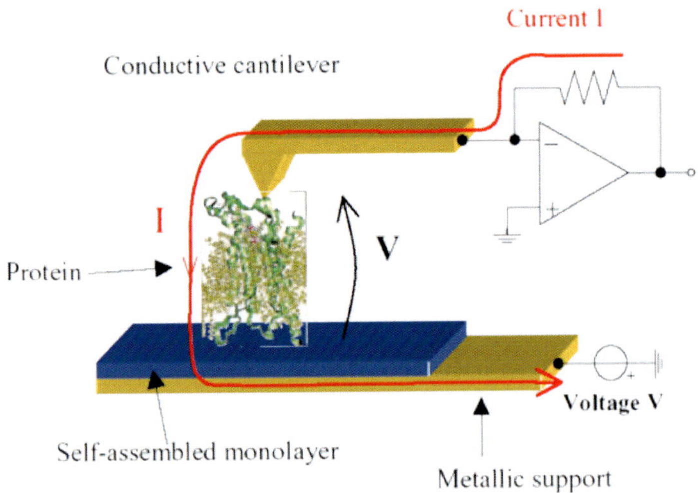


**Figure 3.10** Schematic of the AFM setup including the detection chain. Reprinted from Sampietro et al. (2005) by kind permission of the authors.

[Sampietro et al. (2005)] (with about 1 GΩ gain over about 100 Hz bandwidth), thus allowing DC current measurements in the very wide range $3 \times 10^{-4}$ to 9 nA. The investigations give information on the sample structure and also on its electrical properties. In particular, by using a soft probe, i.e., a probe with a spring constant of 2 N/m [Casuso et al. (2007b)], and a low applied bias (100 mV), it is possible to monitor the sample structure in comparison with its conductivity. In this bias conditions, the PM behaves as an insulator with leakage, i.e., the current levels are lower than 0.3 pA, the setup resolution.

Measurements with higher bias (up to 9 V) should be made by using a stiffer probe (40 N/m), to adsorb part of the electrostatic forces and limiting the sample damages. The acquisition technique is a combination of a dynamic imaging mode and open feedback stepwise approach and allows to test different vertical ($Z$) positions of the tip from air to progressive indentations in the sample [Casuso et al. (2007b)]. With the proposed methodology, the location of the measurement spot in the $XY$ plane is performed using the cantilever controlled in dynamic mode. Once the measuring spot is localized, an open feedback stepwise approach is performed with the cantilever oscillation stopped using the three-dimensional module present in the WSxM software [Horcas et al. (2007)] with a small nominal step (around 0.1 nm). At each step a full wide range of I–V curves is recorded.

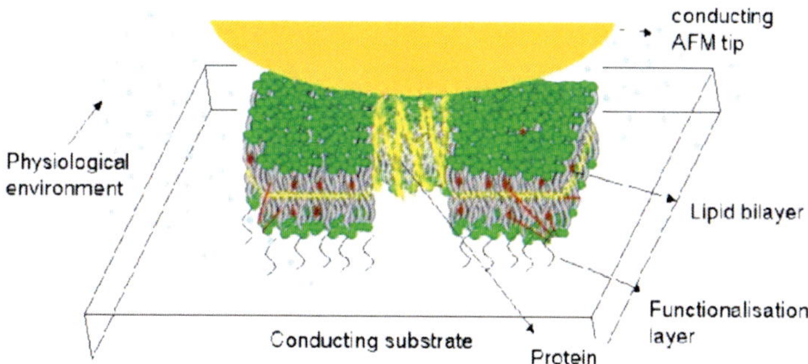


**Figure 3.11** Schematic of the touching between the AFM tip and the single protein.

Figure 3.11 shows a schematic view of the touching between the AFM tip and the single protein.

Simultaneous to electric current, the cantilever deflection is also recorded at each step to monitor the force applied on the sample and to help setting the precise vertical location of the tip. The stiffness of the probe allows for a direct relation between the $Z$ position and the tip-substrate (inter-electrode) distance $L$. After the unavoidable drift of the AFM tip is taken into account, the distance $L$ can be simply obtained, step by step, from the membrane thickness $t_m$ (measured in dynamic mode), the variation in the piezo-displacement $\delta z_p$, and the cantilever deflection $\Delta d$ with respect to the first contact with the membrane. The value of $L$ is then extracted as

$$L = t_m - |\Delta z_p| + |\Delta d| \tag{3.1}$$

The point of first contact with the membrane is easily identified by a reduction in the electrostatic cantilever bending (see discussion below) and a sudden increase in the current flow through the tip at high applied bias. The overall procedure provides the inter-

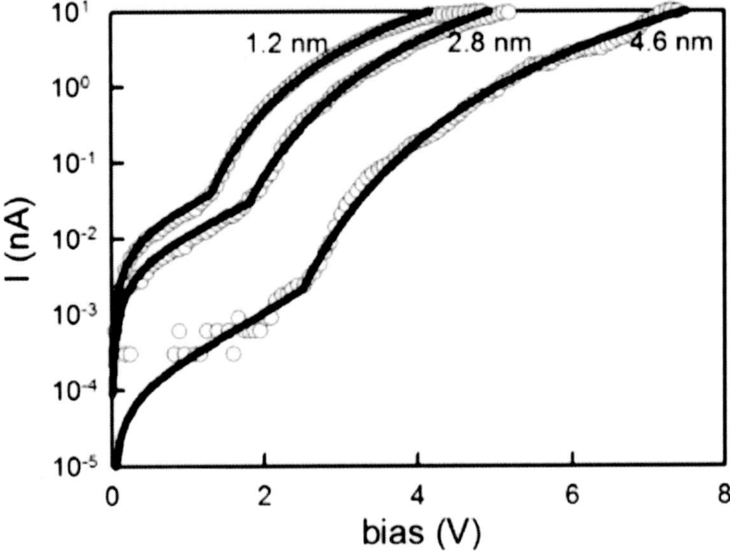


**Figure 3.12** I–V characteristics of bacteriorhodopsin nanolayers in their native state. Tip indentation is reported for each measurement set.

electrode distance with an estimated uncertainty below 0.5 nm. Following this procedure, single point I–V characteristics in a wide voltage range are carried out on various locations of a PM monolayer. Three rounds of measurements on different membranes are usually performed.

Figure 3.12 reports the current–voltage characteristics measured on the PM monolayer at various indentation depths. Here circles refer to measured data and curves to a smoothing interpolation. It is clearly observed that the current flow increases both when the bias is increased and when the inter-electrode distance $L$ is reduced. The current values in the region of minimal PM deformation ($<0.6$ nm) and low bias ($<1$ V) confirm the high resistivity found in the conductive maps for the PM. A detectable current flow (i.e., $I >$ 0.3 pA) is only measured for biases exceeding 1 V, thus yielding a mean resistance at low bias of about 3 TΩ, greater than the 0.3 TΩ lower limit set with conductive maps, thus confirming the contrast previously observed [Casuso et al. (2007b)].

## Chapter 4

# Electrical Properties: Theory

This chapter presents a unified theoretical approach used to interpret protein electrical characteristics and associated properties from a microscopic point of view that considers the three-dimensional amino acids structure.

## 4.1 Theoretical Model

The theoretical model is based on a percolative approach that describes the single protein as a random network of elementary impedances [Akimov et al. (2006); Alfinito et al. (2005); Pennetta et al. (2005)]. By starting with the construction of a time-independent (static) network, corresponding to an ideally frozen protein, all amino acids are taken as single-interacting centers that constitute the backbone three-dimensional structure of the protein [Alfinito et al. (2008, 2009a,b,c, 2010a,b); Alfinito and Reggiani (2009b, 2013); Mukhopadhyay and Lay-Ekuakille (2010)]. The nodes of this static network correspond to the positions of the $\alpha$-carbon atoms of the amino acids, in the native state or in the activated state, as taken from the protein data bank (PDB)

*Proteotronics: Development of Protein-Based Electronics*
Eleonora Alfinito, Jeremy Pousset, and Lino Reggiani

ISBN 978-981-4613-63-7 (Hardcover), 978-981-4613-64-4 (eBook)
www.panstanford.com

[Berman et al. (2000)] or, in its absence, from an homologous modeling [Carloni et al. (2002); Fiser and Sali (2003); Hall et al. (2004); Kitao and Go (1999); Launay et al. (2012a,b); Roy et al. (2010); Sali and Blundell (1990); Vaidehi et al. (2002)]. Then an elementary impedance is associated with each link between a pair of amino acids, established according to a length cut-off criterion. The elementary impedances, which mimic the electrical interactions among the protein amino acids, are taken dependent on the amino acid distance and on other parameters that account for the different physical and chemical properties of the amino acids themselves. By positioning the input and output electrical contacts on the first and the last nodes, respectively, for a given applied bias (current or voltage operation modes according to convenience), the network is solved within a linear Kirchhoff scheme, and its global impedance spectrum, $Z(\omega)$, is calculated in the standard frequency range 0.1–$10^5$ Hz [Alfinito et al. (2008, 2009a,c); Alfinito and Reggiani (2009b)]. When applied to the protein under study, the results of this static, and thus deterministic, network model predict a detectable change in the global impedance associated with the conformational change due to the sensing action. Afterward, the static model is relaxed, and thermal fluctuations associated with the presence of defects and/or with the amino acid oscillations around their average positions are implemented in a dynamic, and thus stochastic, network model. The protein thermal fluctuations lead to an impedance noise, which is calculated as a function of the temperature [Alfinito et al. (2005)].

By construction, besides the small-signal dynamic response, this network produces a parameter-dependent static I–V characteristic determined as [Alfinito et al. (2009b, 2011a)]:

$$V = Z(0)I \tag{4.1}$$

Nonlinearity of the current–voltage characteristics is also considered by implementing the transport mechanism with the inclusion of tunneling processes in the charge transfer between amino acids [Alfinito et al. (2011a,b); Alfinito and Reggiani (2009b)]. To this purpose, an appropriate voltage-dependent resistivity for each link

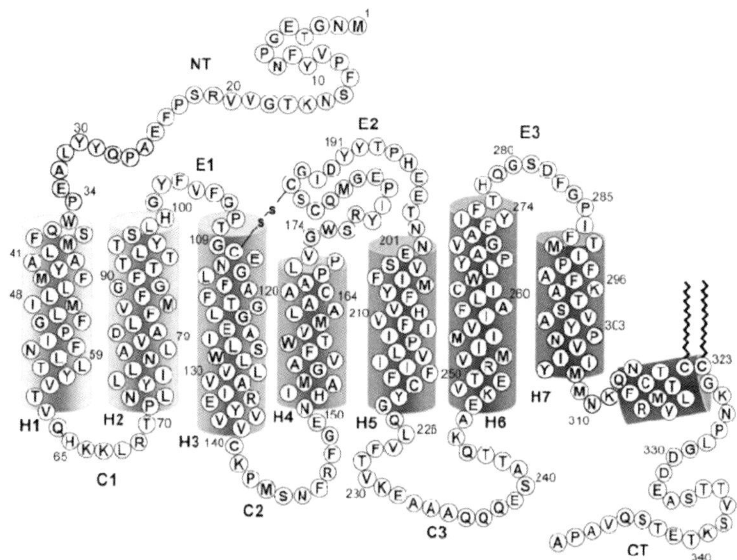


**Figure 4.1** Segment structure of bovine rhodopsin.

is introduced. Further implementations of the model will be detailed at their appearance.

### 4.1.1 *Impedance Random Network*

For illustrative purposes, the light-sensing protein BR belonging to the huge family of G protein–coupled receptors (GPCRs) is taken as a significative prototype.

Figure 4.1 reports the segment (serpentine) structure of BR with letters indicating the amino acids that constitute the protein (see Table 1.1 in Chapter 1 for further details). BR is divided into 15 standard clusters or segments with different functions, structure, and content [Palczewski et al. (2000)]: one *N* terminal (NT), seven transmembrane domains (H1-H7), three intracellular loops (C1-C3), three extracellular loops (E1-E3), and one *C* terminal (CT). The range of segments is indicated in Table 4.1, and Figs. 4.2 and 4.3 report the sphere (left) and backbone (right) representation of bovine rhodopsin in its native and activated (meta) states, as obtained by an homology modeling procedure.

**Table 4.1** Position of bovine rhodopsin segments

| Segment | First residue | Last residue |
|---|---|---|
| NT | 1 | 34 |
| H1 | 35 | 64 |
| C1 | 65 | 70 |
| H2 | 71 | 100 |
| E1 | 101 | 106 |
| H3 | 107 | 139 |
| H2 | 140 | 150 |
| H4 | 151 | 173 |
| E2 | 174 | 199 |
| H5 | 200 | 225 |
| C3 | 226 | 246 |
| H6 | 247 | 277 |
| E3 | 278 | 285 |
| H7 | 286 | 306 |
| CT | 307 | 348 |

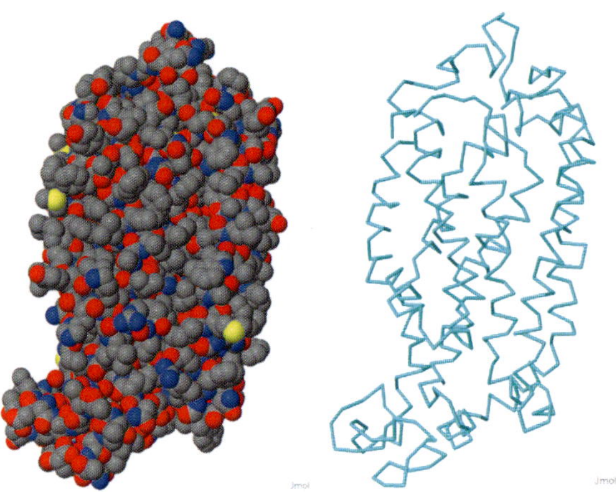

**Figure 4.2** Sphere (left panel) and backbone (right panel) representations of native bovine rhodopsin.

### 4.1.2 *Electrical Properties of a Single Protein*

To evaluate the electrical properties of a single protein, the simple structure presented in Fig. 4.4 is considered. Here the protein

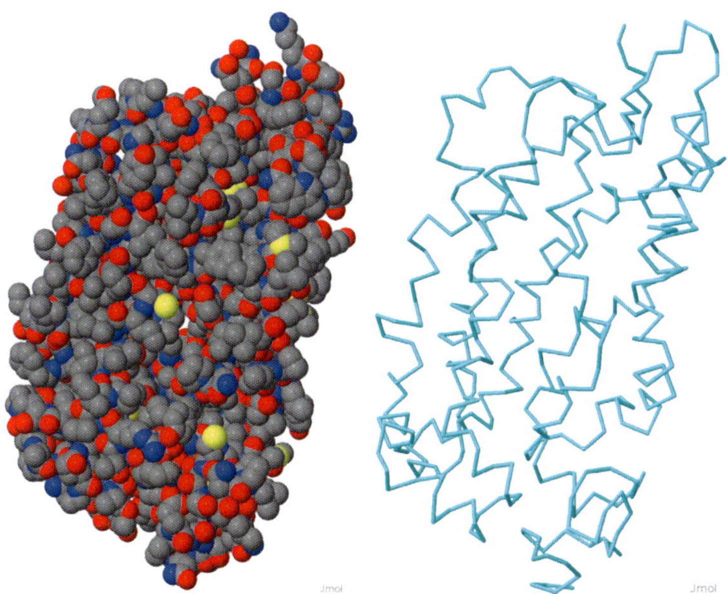


**Figure 4.3** Sphere (left panel) and backbone (right panel) representations of active bovine rhodopsin.

under test is sandwiched between metallic (Ohmic) contacts to which an AC voltage (or current) is applied. Then the device under test (DUT) is modeled as an equivalent circuit consisting of an impedance network. The equivalent circuit can be presented as a simple nondirected graph.

Figure 4.5 reports a visual representation of the graph complexity of BR network (for an interacting radius between neighborhood amino acids $R_a = 5$ Å. The nodes (vertices) of this graph correspond to the amino acids (amino acid residues) of the protein (348 for BR), and the links (edges) between any couple of nodes characterize some kind of interaction (charge transfer and/or charge polarization in the present case) between amino acids, which are neighboring in space within a given radius $R_a$. Two more nodes acting as contacts are then introduced. These contacts are linked to a given set of amino acids (each contact linked to at least one amino acid). Alternatively, the first and the last amino acids of the primary sequence are directly contacted to the external bias.

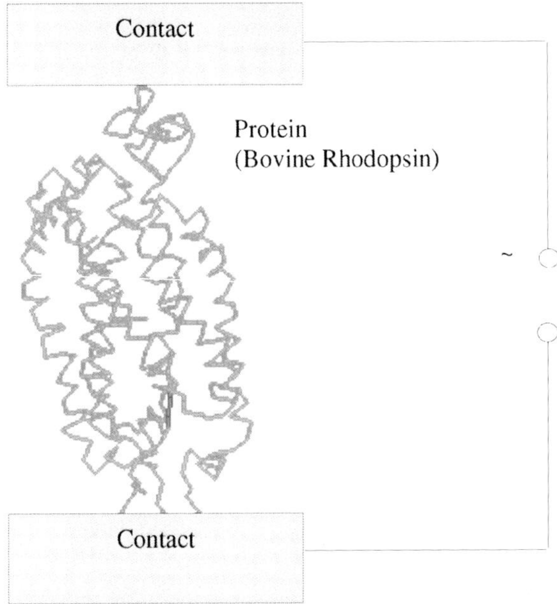


**Figure 4.4** Sketch of the single-protein backbone sandwiched between metallic contacts.

A representation of this interaction network for the case of a hypothetical protein made of 18 residues is shown in Fig. 4.6. One should note that the solution and the membrane are not directly taken into account at this stage of modeling. However, since the main effect of the membrane is to keep the protein in the folded state, this effect is implicitly accounted for by taking the coordinates of the $C_\alpha$ atom corresponding to a given folded conformation of the protein.

The elementary impedance of a link is taken as the most usual passive equivalent AC circuit made of a resistor $R$ in parallel with a parallel-plate capacitor $C$. To construct the impedance network, the amino acids are considered uniform spheres with the center at the $\alpha$-carbon atom and a typical van der Waals radius $R_a = 5$ Å. When two spheres overlap (i.e., the distance between their centers $l_{i,j} < 2R_a$, $i$ and $j$ being the indexes of the two considered amino acids), they are assumed to be neighboring and the impedance edge should be inserted between the corresponding nodes of the

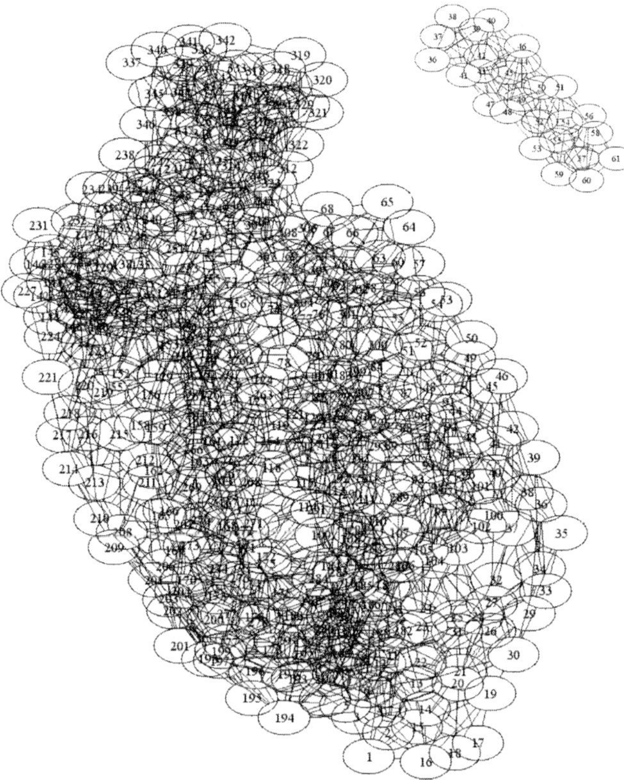


**Figure 4.5** Representation of BR network graph ($R_a = 5$ Å). The picture shows the network topology without any correspondence with spatial coordinates. The numbers inside the circles represent the label of the residues. The graph of the first transmembrane domain H1 constructed by the same rules is shown separately to evidence its structure.

circuit (see Fig. 4.7). Otherwise, there is no interaction between amino acids and thus no link between the corresponding nodes. The determination of the value of each elementary impedance can be fixed with increasing degree of complexity. Here three different approaches are considered. (i) In the first approach, the elementary impedances of the edges, $Z_{i,j}$, are taken to be all the same. (ii) In the second approach, $Z_{i,j}$ is taken to be proportional to $l_{i,j}$ as for a simple Ohmic resistor and/or a planar homogeneous capacitor. In

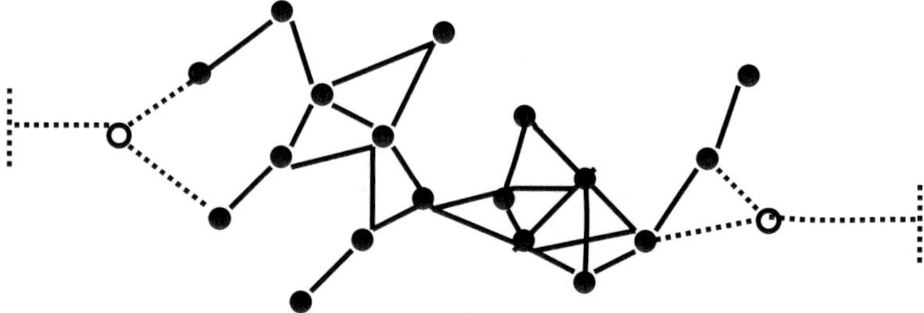

**Figure 4.6** Interaction network associated with a hypothetical protein made of 18 residues: The full circles show the nodes positioned at the $\alpha$-carbon atom of each amino acid, and the lines represent the links arising from electrical interactions between a pair of amino acids with relative distance shorter than a cut-off value. The open circles represent two extra nodes associated with the electrodes.

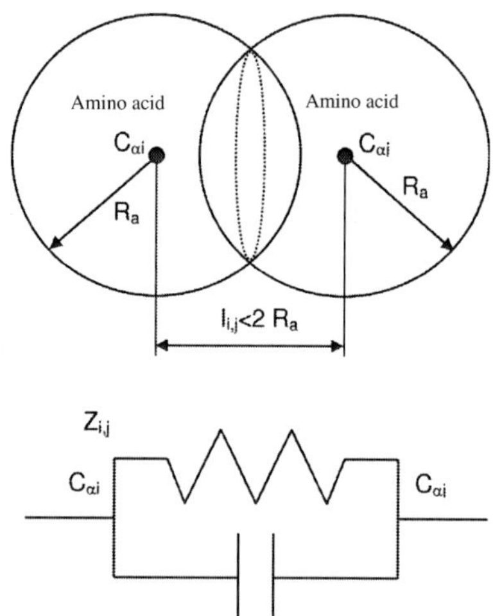


**Figure 4.7** Overlap between two amino acids and equivalent circuit element. $C_i$ identifies the center of the sphere corresponding to the $\alpha$-carbon atom of the $i$-th amino acid.

this case, the link impedance is given by:

$$Z_{i,j} = \frac{l_{i,j}}{A} \frac{1}{(\rho^{-1} + i\epsilon_{i,j}\,\epsilon_0\omega)} \tag{4.2}$$

where $A$ is the cross-sectional area of the capacitor and of the resistor; $\rho$ is the resistivity, taken to be the same for every amino acid, with the indicative value of $\rho = 10^{10}$ $\Omega$ m; $i = \sqrt{-1}$ is the imaginary unit; $\epsilon_0$ is the vacuum permittivity; and $\omega$ is the circular frequency of the applied bias.

The relative dielectric constant of the couple of $i$, $j$ amino acids forming the corresponding link, $\epsilon_{i,j}$, is expressed in terms of the intrinsic polarizability of the $i$, $j$ amino acids [Song (2002)] as:

$$\epsilon_{i,j} = 1 + [(\alpha_i + \alpha_j)/2 - 1] \times 79/17 \tag{4.3}$$

where $\alpha_i$ and $\alpha_j$ are the intrinsic polarizabilities of the corresponding amino acids taken from reference [Song (2002)], as already reported in Table 1.1. This expression is constructed to distribute $\epsilon_{i,j}$ between 1 and 80 (vacuum and water) proportionally to $(\alpha_i + \alpha_j)/2$. (iii) In the third approach, by assuming that the cross-sectional area of the resistor and the capacitor is equal to the cross-sectional area of the overlapping spheres, $\mathcal{A}_{i,j} = \pi(R_a^2 - l_{i,j}^2/4)$ is taken as the cross-sectional area between two spheres of radius $R_a$ centered at the $i$-th and $j$-th nodes, respectively. The final expression takes the form:

$$Z_{i,j} = \frac{l_{i,j}}{\mathcal{A}_{i,j}} \frac{1}{(\rho^{-1} + i\epsilon_{i,j}\,\epsilon_0\omega)} \tag{4.4}$$

Further approaches can be introduced to account for the properties of real amino acids, for example, polarizability, chemical bonds between amino acids, etc.

In going from approach (i) to approach (iii), a better sensitivity of the network to the change of its structure is exploited. To this purpose, Fig. 4.8 reports the behavior of the absolute value of the network impedance $|Z|$ versus $R_a$ for these three models. The systematic decrease in $|Z|$ at increasing $R_a$ reflects the increasing importance of parallel with respect to series connections. The continuous increase in $R_a$ is equivalent to a continuous three-dimensional contraction of the structure and represents an example of a spatial conformation dependence of $|Z|$. One can see that the curves of the first two models exhibit a step-like shape related to a sharp discontinuity of the value of $|Z|$ when $R_a$ becomes equal to $l_{i,j}$, while the third model exhibits a smooth behavior. Everywhere below, the third model is used by default.

The input data necessary for describing the graph are constructed from the standard protein data bank (PDB) file (see the next

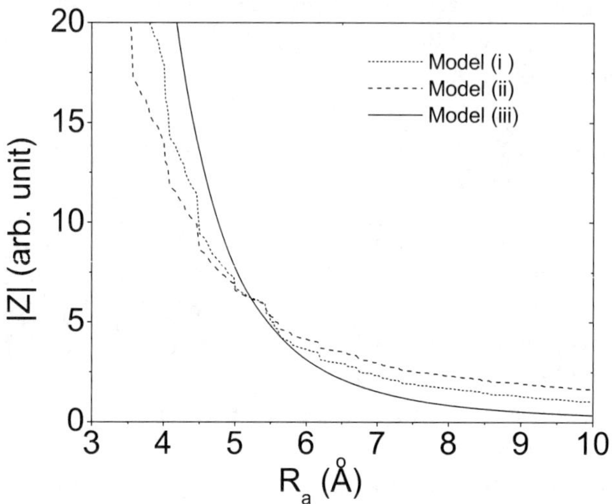


**Figure 4.8** Comparison among the three models of calculating the single-protein impedance.

section). The obtained graph should be checked to identify defects as disconnected graph and pendant branches, which can cause fatal error during numerical processing.

The model so developed is hereafter called impedance network protein analogue. This one-node impedance model, henceforth also called AA model, is further implemented by considering also the case of having two nodes for each amino acid, henceforth also called AB model. Indeed, by identifying the amino acid with its $C_\alpha$ in the one-node model, the resulting structure becomes analogous to the polypeptidic backbone. However, in this way only the backbone behavior can be reproduced, while in a conformational change, the backbone displacement is not the only relevant transformation. Actually, rotations of each amino acid around the backbone [Sheu (2002)] can also affect the electrical properties of the protein. To account for these possibilities, one should depart from the one-node model and look for a more realistic picture. Since the distinctive mark of each amino acid is in its residue, which does not lie on the polypeptidic backbone, it is natural to fix on each amino acid a second node. Accordingly, as the second node is chosen the so-called

$C_\beta$ atom, i.e., the second carbon atom that attaches to the functional group [Lattanzi (2002)]. The $C_\beta$ atom is present in all the amino acids with the exclusion of glycine. The impedance attributed to the new links arising from the presence of this second node is taken of the same form as in Eq. (4.4). In this way, the total number of nodes mapping the protein, $\nu$, is practically doubled. To connect the nodes, two choices are adopted: the isotropic network and the directed network; the latter choice being able to better exploit directional characteristics in analogy with the directed percolation [Ódor and Szolnoki (1996)]. In the isotropic network, the $\alpha$ and $\beta$ nodes are considered to be equivalent. Accordingly, each node is connected with all the others inside the interaction radius. Thus, for sufficiently large $R_c$, each node has $(\nu - 1)$ connections. In the directed network, the $\alpha$ and $\beta$ nodes are not equivalent. Accordingly, each $\alpha$ node, identified by the serial number of the protein primary structure, is connected with all the $\alpha$ nodes inside the interaction radius. The same happens for the links between $\beta$ nodes. By contrast, each $\alpha$ node is linked to $\beta$ nodes pertaining to amino acids with higher or equal serial number. One should notice that for the isotropic network, the maximum value of the total number of links is:

$$N^{max}_{isotropic} = (N_\alpha + N_\beta)(N_\alpha + N_\beta - 1)/2, \tag{4.5}$$

while for the directed network, it is:

$$N^{max}_{directed} = (N_\alpha)(N_\alpha - 1)/2 + (N_\beta)(N_\beta - 1)/2 + \sum_{\alpha} \sum_{\beta \geq \alpha} N_\alpha N_\beta. \tag{4.6}$$

Here $N_\alpha$ is the number of $C_\alpha$, coincident with the total number of amino acids, $N$, and $N_\beta$ is the number of $C_\beta$. Since BR contains 348 $C_\alpha$ atoms and 325 $C_\beta$ atoms, one can assume $N_\beta \approx N$ and then deduce the expressions:

$$\begin{aligned} N^{max}_{isotropic} &\approx 2N(2N - 1)/2, \\ N^{max}_{directed} &\approx N(3N - 1)/2. \end{aligned} \tag{4.7}$$

By comparing Eq. (4.5), Eq. (4.6), and Eq. (4.7), one notices a global different functioning of the isotropic and directed networks. In the former network, each node represents an independent unity that interacts in the same way with all the other unities by elongating $(N_\alpha + N_\beta - 1) \approx 2N - 1$ links. In the latter network, each amino acid represents a working unit that interacts with other units through

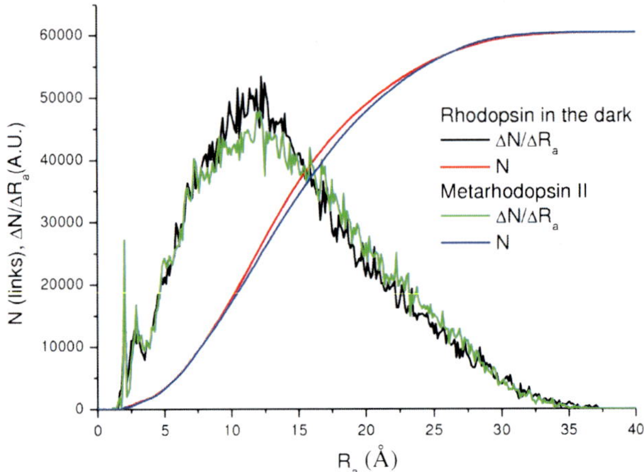


**Figure 4.9** Number of links $N$ and its increment $N/R_a$ versus $R_a$ for BR in dark and metarhodopsin II impedance network. $R_a < 2$ Å, then $N < 348 \rightarrow 0$ disconnected network; $R_a \approx 2$ Å, then $N \approx 348 \rightarrow$ sequential limit; $R_a > 35$ Å, then $N = (348 \times 347)/2 = 60378 \rightarrow$ full network.

only three kinds of links, respectively: $C_\alpha$-$C_\alpha$, $C_\alpha$-$C_\beta$, and $C_\beta$-$C_\beta$. The number of links drawn out by each amino acid is $\approx 3N - 1$. One should notice that the isotropic network exhibits a number of links that is in general larger than the directed network up to a maximum value of about 25%.

In what concerns with the contacts to the external bias, in the AA model these are usually positioned on the first and the last amino acids. In the AB model, three possibilities are explored. Accordingly, one should take as injector and collector nodes: (i) the first $C_\alpha$ and the last $C_\beta$, $AB_{\alpha,\beta}$ model; (ii) the couple $\alpha - \beta$ of the first amino acid, and the couple $\alpha - \beta$ of the last amino acid, $AB_{\alpha\beta,\alpha\beta}$ model; (iii) the first $C_\alpha$ atom and the last $C_\alpha$ atom, $AB_{\alpha,\alpha}$ model.

### 4.1.3 *Network Properties of the Protein Under Test*

The network properties of a protein are analyzed in terms of the number of links and its increment as a function of the interaction radius.

Figures 4.9–4.12 report the degree distributions for different values of the interaction radius and for three GPCRs.

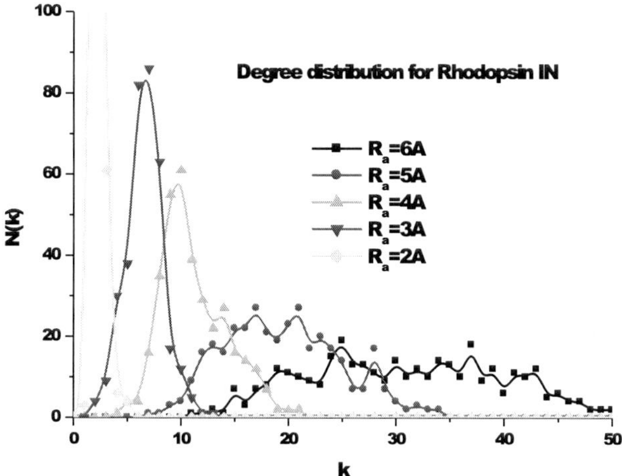


**Figure 4.10** Degree distribution for the bovine rhodopsin impedance network. Splined line and scatter graph for the different $R_a$ in the same scale. Inset shows bar graph for the case of $R_a = 5$ Å.

### 4.1.4 *Calculation of a Single-Protein Molecular Volume*

Here the molecular volume of a single protein is calculated as the total volume occupied by the crossing spheres with a characteristic van der Waals radius $R_a = 5$ Å around the $\alpha$-carbon atom of each amino acid. To this purpose, the following Monte Carlo (MC) technique is used. For each amino acid in the cycled series, randomly select a point inside the sphere with relative polar coordinates $r_i = r_{1,i} R_a$; $\theta = 2\pi r_{2,i}$; $\phi = \pi r_{3,i}$ with $r_{j,i}$ a random number evenly distributed between 0 and 1 corresponding to the $i$-th iteration. The corresponding elementary volume is $\Delta V_i \sim r_i^2 \sin\theta_i$. The molecular volume is then found as:

$$V = V_{tot} \lim_{M\to\infty} \frac{\sum_{i=1}^{M}(\Delta V_i / N_i)}{\sum_{i=1}^{M} \Delta V_i} \tag{4.8}$$

where $V_{tot} = n \times 4/3\pi R_a^3$, $n$ is the number of amino acids (348 for BR), $N_i$ is the number of spheres to which the $i$-th random point belongs to.

Calculations predict a molecular volume of BR in dark of 60.6 $nm^3$ (MC over 13.880.000 points), a volume of BR in light of 63.9 $nm^3$ (MC over 20.610.000 points). Thus, the BR volume for the

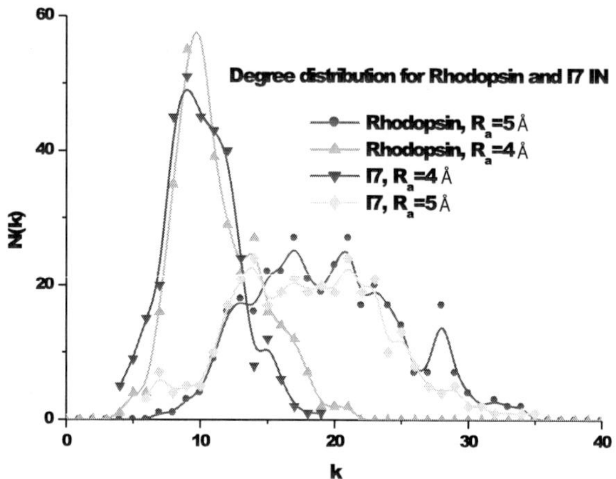


**Figure 4.11** Degree distribution for bovine rhodopsin and rat OR I7 impedance network. Splined line and scatter graph for the different $R_a$ in the same scale.

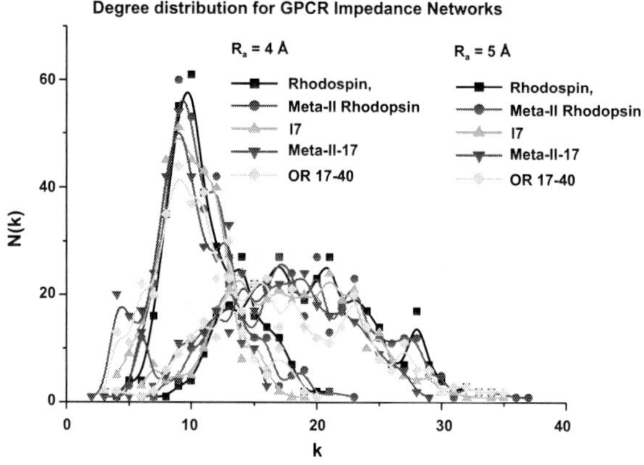


**Figure 4.12** Degree distribution for the different GPCR considered. Splined line and scatter graph for the different $R_a$ in the same scale.

activated state is of about 5.4% greater than that for the native state, and practically independent of the value of $R_a$. Figure 4.13 reports the percent variation of the volume of the BR associated with its conformational change as function of the interacting radius.

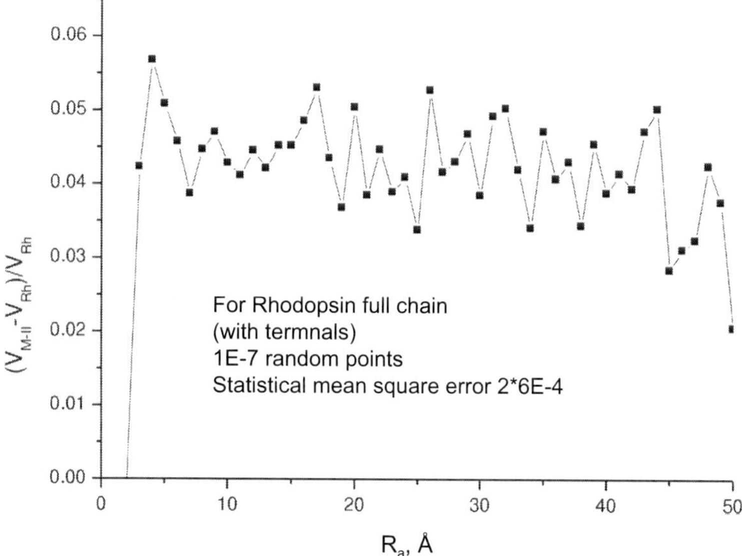


**Figure 4.13** Differences between activated and ground state volumes of the bovine rhodopsin single molecule with terminals.

## 4.1.5 *Conformational Process: General*

Here the objective is to first develop an approach to the modeling of the process of the BR molecule conformational change from the native state (BR in dark) to the bovine metarhodopsin II activated state (BR in the presence of light). To this purpose, two models that describe the continuous transformation are developed.

In the first model (coordinate model), the three coordinates of each amino acid of BR in dark are linearly transformed into the three coordinates of bovine metarhodopsin II. The initial and final configurations are monitored in 100 steps.

In the second model (length model), the lengths between each pairs of amino acids, $(348 \times 347)/2 = 60378$, are linearly transformed from the configuration of BR in dark to bovine metarhodopsin II. Even in this case, the initial and final configurations are monitored in 100 steps.

### 4.1.6 *Conformational Process: Coordinate Model*

In the first model, the initial and final states are superimposed by using the following procedure. First, one selects three amino acids A, B, and C as basis, with coordinates $\mathbf{R_A}$, $\mathbf{R_B}$, $\mathbf{R_C}$ for the native state and $\mathbf{R'_A}$, $\mathbf{R'_B}$, $\mathbf{R'_C}$ for the activated state. Second, each structure is transferred by isomorphic motion and rotation to the new coordinate system, in which (i) $\mathbf{R_A} = \mathbf{R'_A}$, (ii) the vectors **A**, **B** are parallel: i.e., $(\mathbf{R_B} - \mathbf{R_A})||(\mathbf{R'_B} - \mathbf{R'_A})$, and (iii) the planes $(ABC)$ coincide: i.e., $(\mathbf{R_A}, \mathbf{R_B}, \mathbf{R_C}) = (\mathbf{R'_A}, \mathbf{R'_B}, \mathbf{R'_C})$. Third, one selects a number of intermediate states that describe the transition from the native to the activated state, where the coordinates of each $i$-th amino acid are determined as:

$$\mathbf{R}'_i(f) = \mathbf{R}'_i + f(\mathbf{R}'_i - \mathbf{R}_i) \tag{4.9}$$

where $\mathbf{R}_i$, $\mathbf{R}'_i$ are the coordinates of corresponding amino acids of the native and activated states, and $f \div (0 - 1)$ is the fraction of linear conformation.

For $f = 0$ and $f = 1$, the structures of the native and activated states are recovered, respectively. For each intermediate state, which is assumed to describe the transition between the native and activated states, the corresponding network is constructed and the impedance calculated. The following figures report the results of such a modeling. Here, the basis of amino acids is taken at numbers 40, 155, 282. These are the points used for the construction of the full PDB data and correspond to the minimum average deviation of the different BR PDB data.

Figure 4.14 reports the change in the molecular volume when going from the native to the activated state of BR. The volume exhibits a minimum value around $f = 0.5$ and a final net increase for about 5.4% in agreement with the results reported in Fig. 4.13. In particular, the volume of the BR molecule in dark is 60.6 $\text{nm}^3$ (MC runs over $13 \times 10^6$ points), and the volume of the bovine metarhodopsin II molecule is 63.9 $\text{nm}^3$ (MC runs over $20 \times 10^6$ points).

Analogous simulations are performed starting from different sets of basis points. Figure 4.16 reports the results of the relative change in impedance as a function of the linear conformation parameter $f$ for the different models of elementary impedance. Figure 4.17

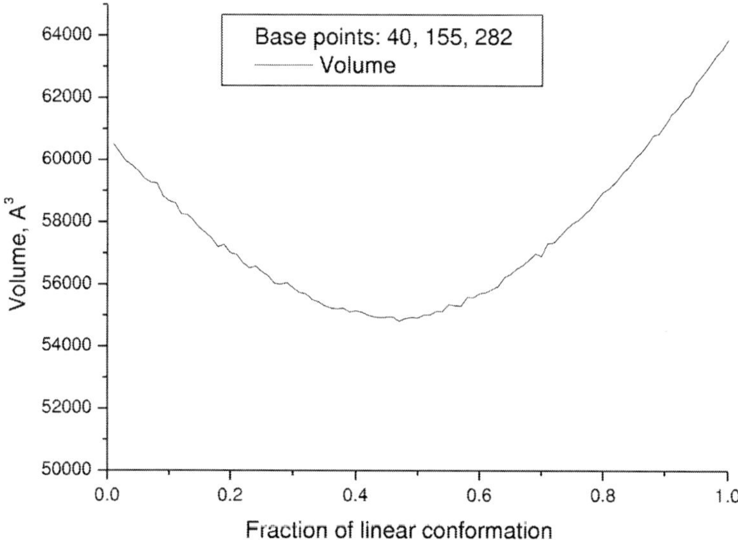


**Figure 4.14** Molecular volume versus fraction of linear conformation for the first conformation model.

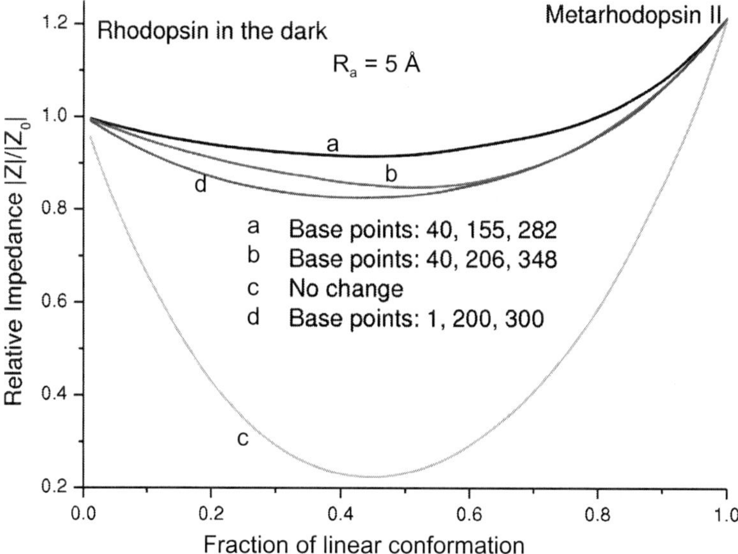


**Figure 4.15** Relative impedance versus fraction of linear conformation, first conformation model. The curve "no change" is obtained by applying the linear conformation directly to the two structures as given in PDB files.

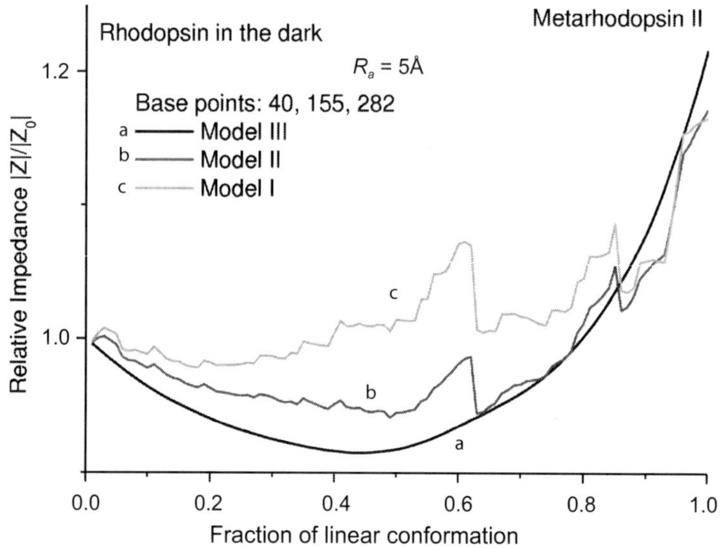


**Figure 4.16** Relative impedance versus fraction of linear conformation for the different impedance models reported in the figure, first conformation model.

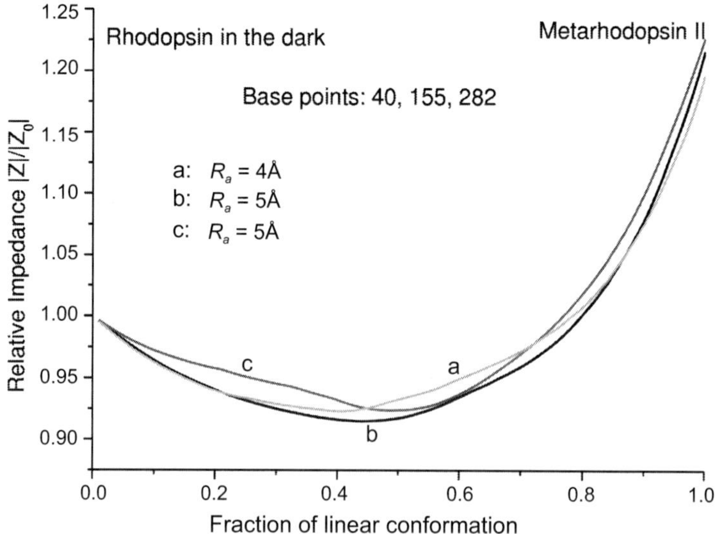


**Figure 4.17** Relative impedance versus fraction of linear conformation for the different interacting radius reported in the figure, first conformation model.

reports the results of the relative change in impedance as a function of the linear conformation parameter $f$ for different values of $R_a$.

The value of the single-protein impedance of the modeled intermediate states for the first linear conformation model exhibits a minimum for all the choices made here of the three amino acids used as basis as well as for a given basis for all the values of $R_a$. The minimum is a property related to the structure and to the linear transformation assumed here. The selected set of basis points is rather significant for the result, but the minimum persists independently of the selected set. It can be explained by the fact that in a three-dimensional space, during a linear conformation into each other, randomly oriented segments pass in most cases through a minimum value of length.

### 4.1.7 *Conformational Process: Length Model*

In the second model, the state of the molecule structure is described as an array of lengths between each pair of amino acids: $l_{i,j} = |\mathbf{R}_i - \mathbf{R}_j|$, $i, j = 1 \ldots N$, where $N$ is the number of amino acids. On the basis of this array, data for the calculations can be generated. Accordingly, the array for the intermediate steps is taken as:

$$l''_{i,j}(f) = l_{i,j} + f(l'_{i,j} - l_{i,j}) \tag{4.10}$$

where $l_{i,j}$ and $l'_{i,j}$ correspond to the native and activated states of bovine rhodopsin. Note that the number of links changes during the transformation according to the comparison of $l_{i,j}$ with $R_a$. This model is simpler than the first one; it does not need additional geometrical manipulations and does not require additional parameters such as the basis points. However, it is less intuitive since the described states are characterized by $(N^2 - N)$ parameters, while the real three-dimensional structure of $N > 3$ nodes has only $3N - 5$ internal degrees of freedom. It means that the intermediate states obtained in such a way are "virtual"; they have no correct analogues in a real three-dimensional geometry, while their impedances can still be calculated.

The results of the length conformational model are reported in Figs. 4.18 and 4.19. Here the related impedance modulus exhibits a maximum for small $R_a$ and a linear increase for large $R_a$.

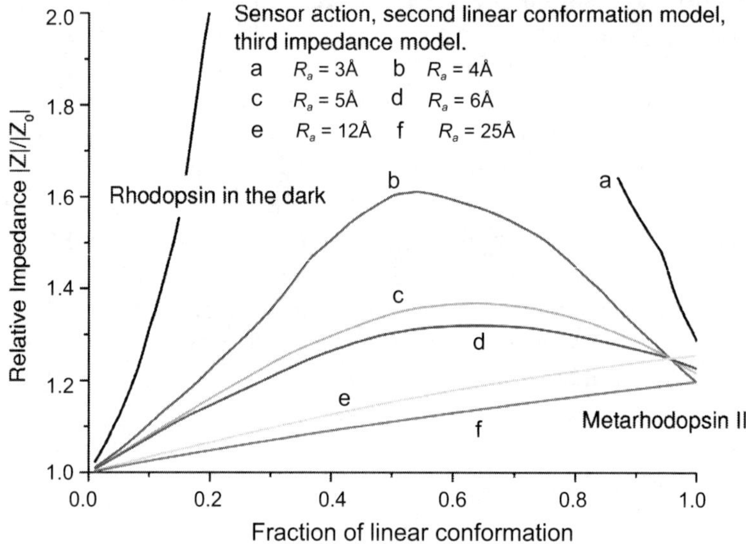


**Figure 4.18** Relative impedance versus fraction of linear conformation for the different interacting radius reported in the figure, second conformation model.

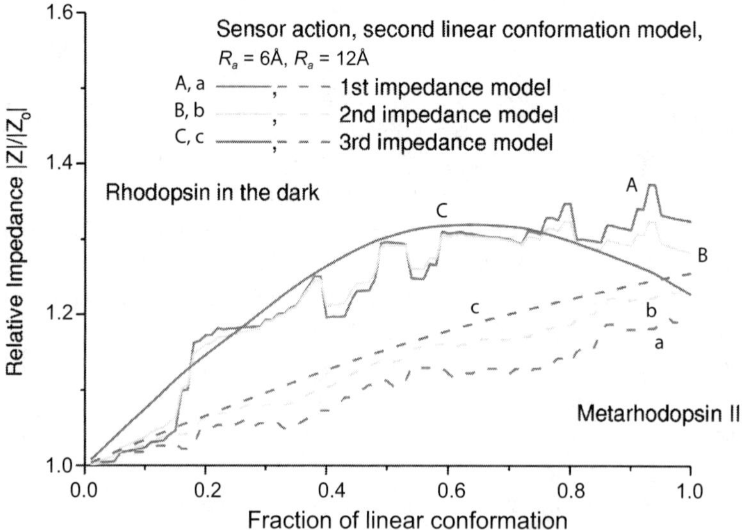


**Figure 4.19** Relative impedance versus fraction of linear conformation for the different impedance models reported in the figure, second conformation model.

These findings can be explained in the following way. If during the conformation, the amplitude of the mutual motion of nodes is much greater than $R_a$, then links arise in intermediate states, which do not exist in both the starting and ending states of the structure. Such intermediate links are not taken into account in the second model. Accordingly, for the intermediate states, the calculated number of links is less than the real one, thus implying an increase in resistance.

### 4.1.8 *Topological Investigation*

By using the so-called contact maps of a network, this section reports some results that constitute a two-dimensional map of the network connected parts, and that can be drawn by assigning to each couple of linked amino acids, say $i$, $j$, a point of coordinates $i$, $j$. Two amino acids are linked when their distance is less than the interaction radius $R_a$; otherwise, they are not linked. Therefore, by comparing the contact maps of the protein in the native and activated states, one can have access to a direct view of the main changes that are induced by the sensing action in the protein topology.

Figures 4.20 and 4.21 report the contact maps of rat OR I7 calculated with the two different $R_c$ values of 6 and 12 Å, respectively [Alfinito et al. (2011b)].

From the above figures, one can observe that by increasing the value of $R_c$, the density of points, and thus the connectivity, increases substantially, as expected. In particular, the main differences between the native and activated states are in the increased connectivities among the helices h3-h4-h6, where the principal binding pocket is presumably located [Alfinito et al. (2011b); Hall et al. (2004)]. Notably, significant differences between the native and activated states survive for large $R_c$, say up to $R_c$ values of about 50 Å. A quite similar behavior has been observed for other GPCRs [Alfinito et al. (2011b)].

### 4.1.9 *Resistance and Impedance Spectrum*

The block diagram with the details of the code used to calculate the resistance and the impedance spectrum of a single protein is

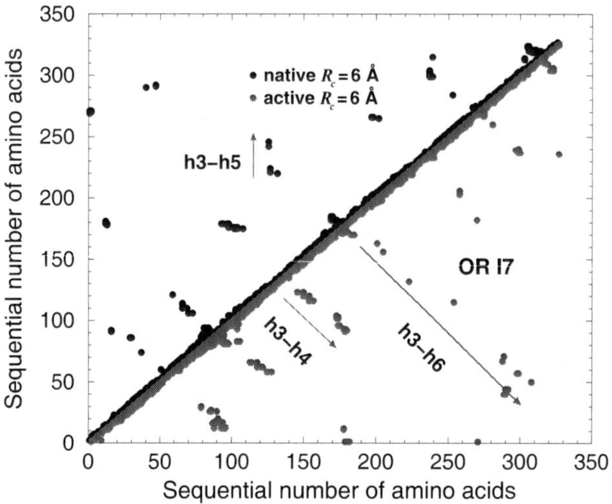


**Figure 4.20** Contact map of the two rat OR I7 representations, native state (squares on the left), and activated state (squares on the right). Because of the axial symmetry with respect to the diagonal exhibited by data, the symbols pertaining to the native map are reported only on the left-hand side of the diagonal and those pertaining to the activated map on the right-hand side. The interaction radius is $R_c = 6$ Å.

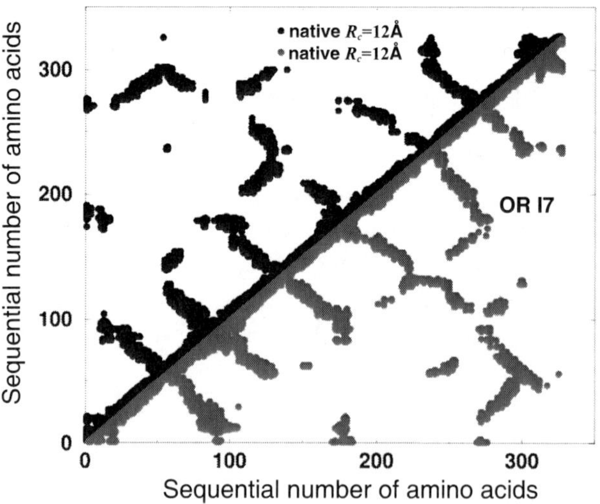


**Figure 4.21** Same as in Fig. 4.20 for an interaction radius of $R_c = 12$ Å.

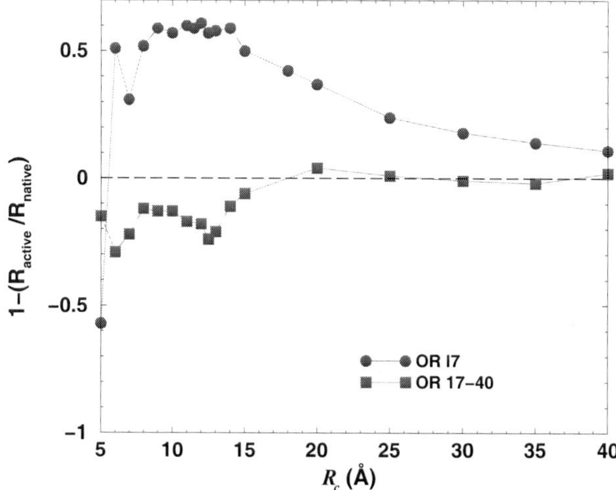


**Figure 4.22** Relative resistance variation for increasing $R_c$ values. Circles refer to rat OR I7, and squares refer to human OR 17–40.

reported in the Appendix. Some proteins of interest are illustrated below the main results obtained by the theory.

The evidence of the differences between the native and activated states of a given protein is found in the change of the receptor global resistance. Accordingly, Fig. 4.22 reports the dependence of the relative variation of resistance, $(1 - R_{active}/R_{native})$, as a function of $R_c$, for two ORs belonging to the GPCR family, rat OR I7 and human OR 17–40. The main result reported by this figure is the larger sensitivity exhibited by rat OR I7 with respect to that of human OR 17–40. In the former case, it is possible to resolve a maximum difference of about 60%, while in the latter case, the maximum resolution is only about 20%. The region of maximum sensitivity is the same for both the proteins and corresponds to $R_c$ in the range 6–14 Å. For the case of human OR 17–40, the activated state takes resistance values that are significantly smaller than those of rat OR I7. Furthermore, for $R_c$ above about 18 Å, calculations provide evidence for human OR 17–40 of an inversion of the resistance variation, with the activated state becoming less resistive than the native state in two regions of $R_c$ values, 18–24 Å and 40–60 Å, respectively. In terms of the electrical network, such an inversion is

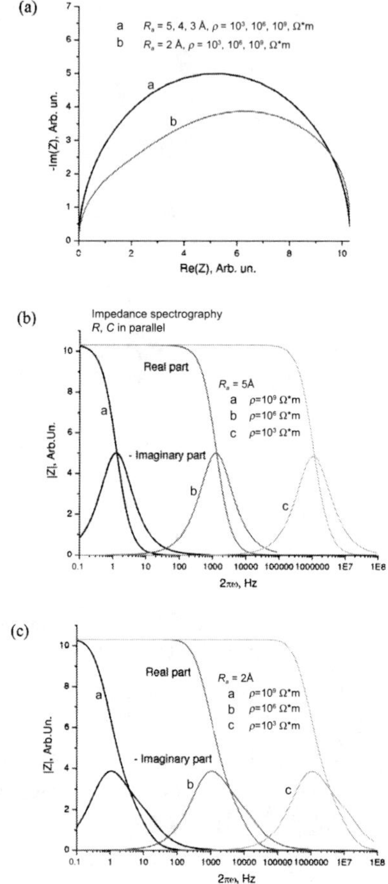


**Figure 4.23** Impedance spectroscopy of the considered impedance network that models bovine rhodopsin. Impedance values are normalized to the maximum of the real part.

interpreted as a stronger increase in parallel with respect to series connections. Accordingly, the different behavior shown by the two receptors signals different peculiarities of the network structure.

The impedance spectrum of single proteins is explored on a wide range of frequencies and the results are given by means of the Nyquist plot and/or the Bode plot, as already reported in Section 3.2.3. To this purpose, Fig. 4.23 reports the Nyquist plot (part a) and the Bode plots (parts b and c) of impedance networks built with the procedure described above, starting from the BR structure. As a general trend, the shape of the Nyquist spectrum is undistinguishable from that obtained from a single impedance, except for the cases with $R_a = 2$ Å, when the degree of most of the nodes of the graph is equal to 2 (see Fig. 4.10) and the series combination of elementary impedances is predominant in the impedance network structure. Impedance networks with predominance of parallel connection of RC elements always yield a semicircular shape on the Nyquist plot. Changes in shape are only detected in the case of the sequential limit and/or in the presence of strongly different values of the link dielectric constants.

### 4.1.10 *Random Fluctuations in the Impedance Network*

The introduction of stochastic fluctuations in the impedance network is considered below. This task can be pursued by allowing different mechanisms of stochasticity. At present, the following approach is adopted. Instead of the static network described in the previous sections (also called perfect and/or frozen network), a fluctuating network, where some links can be randomly broken and recovered, is considered. Accordingly, one can define two probabilities, $W_b$ and $W_r$, that represent the breaking and the recovery probability for each link, respectively. Each configuration (state) of the random network is then characterized by the fraction of broken links, $p$.

The value of $p$ at which the network breaks (i.e., it becomes an open circuit), $p_c$, is a characteristic value (percolation threshold) that depends on the network topology. The block diagram of the algorithm used to find the value of $p_c$ is reported in the Appendix.

The histogram of the distribution of $p_c$ at the percolation threshold for an impedance network modeled on BR is shown in the left panel of Fig. 4.24. Here the asymmetry of the distribution is due to a value of the percolation threshold close to one for the chosen values of the parameters. As a validation check of the used algorithm, the same simulations are carried out for a regular square network $20 \times 20$ (see the right panel of Fig. 4.24) and they are found

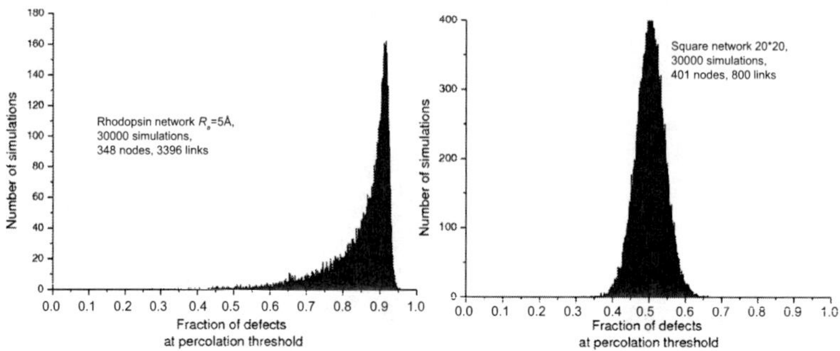


**Figure 4.24** Distribution of the fraction of defects at percolation threshold.

to confirm the result predicted by theory: $p_c = 0.5$. The dependence of $p_c$ on $R_a$ for a BR network is reported in Fig. 4.25.

The Monte Carlo simulator is further implemented to calculate the evolution of a random impedance network when the two mechanisms of breaking and recovery, previously described, are

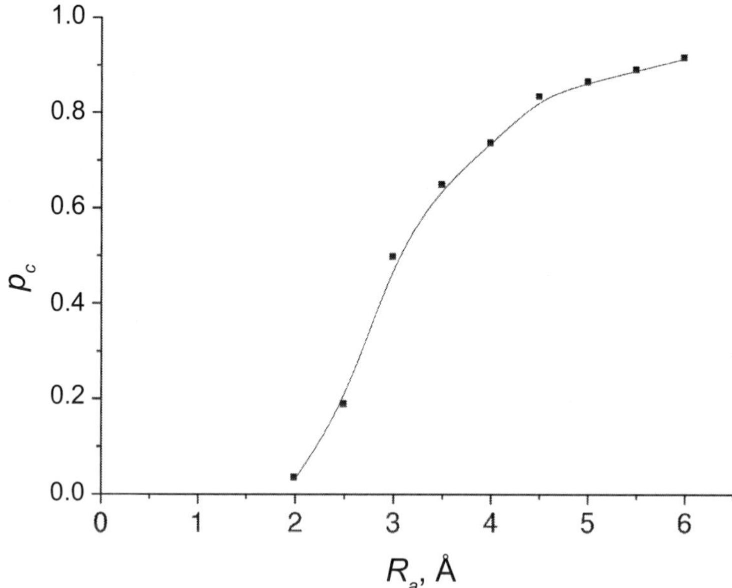


**Figure 4.25** Average fraction of broken link at threshold versus $R_a$ for the bovine rhodopsin network. Each point is averaged over 30,000 realizations.

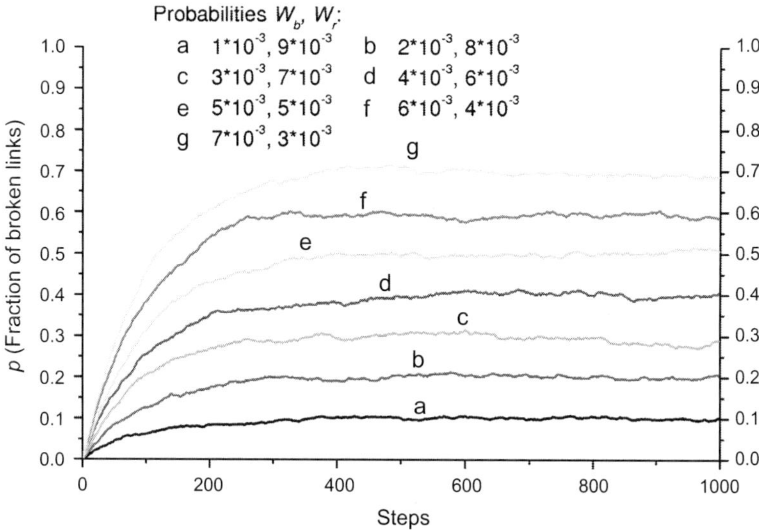


**Figure 4.26** Fraction of broken links in bovine rhodopsin network ($R_a = 5$ Å) during MC simulation with different $W_0b$, $W_r$.

active. In this case, if a link is broken, then the nominal value of the corresponding resistivity $\rho$ is multiplied by $10^8$ and $\epsilon$ is divided by the same value, which practically corresponds to an infinite value of the link impedance. Then, according to Eq. (4.4), the imaginary and real parts of the corresponding elementary impedance are changed by the same quantity.

Figure 4.26 reports the fraction of broken links $p$ within the first 1000 iteration steps for different combinations of $W_b$ and $W_r$ values. The evolution curves show that similar to the case of regular random networks [Pennetta et al. (2000)], the impedance network reaches a steady state characterized by an average value of broken links. It is possible to show [Pennetta et al. (2000)] that this average value is completely determined by the values of $W_b$ and $W_r$.

Figure 4.27a shows the absolute value of the network impedance with $R_a = 5$ Å, $\rho = 16\ \Omega$m, $2\pi\omega = 50$ Hz), while Fig. 4.27b displays the real and negative imaginary parts of the network impedance.

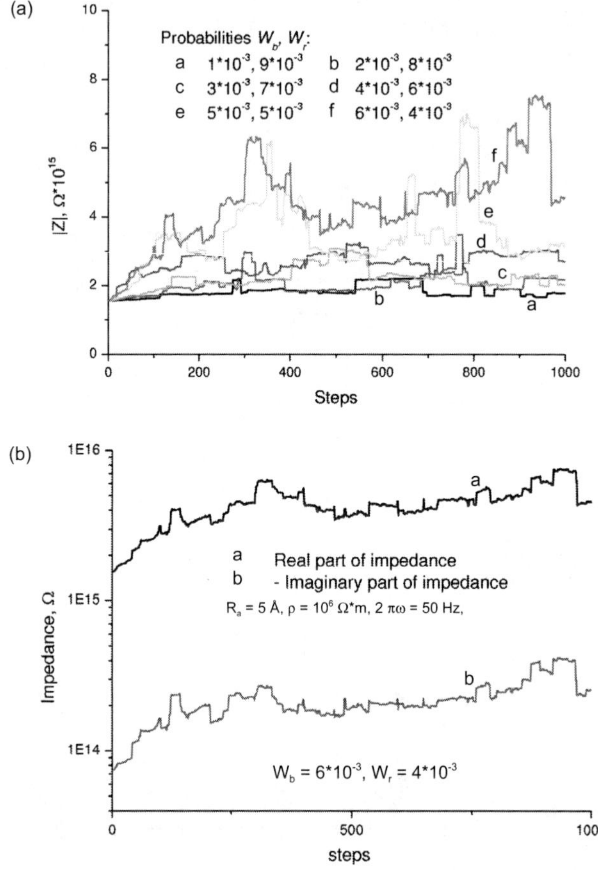


**Figure 4.27** (a) Absolute value of the impedance of bovine rhodopsin network ($R_a = 5$ Å, $\rho = 10^{16}$ Ωm, $f = \omega/2\pi = 50$ Hz) during MC simulation with different $W_b$, $W_r$. (b) Real and negative imaginary parts of total network impedance during MC simulations.

## 4.2 Dynamic Fluctuations of the Impedance Network: Oscillator Models

In this section, the frozen protein approximation is relaxed and the thermal fluctuations of the amino acids' $C_\alpha$ position around its average values are included in the model. To this purpose, a classical harmonic-oscillator approach is first used, and then the case of a quantum harmonic-oscillator approach is considered. Accordingly, the first approach adopts a single-force constant model [Alfinito et al. (2005); Bahar (1997); Pennetta et al. (2005); Tirion (1996)] as the simplest level of modelization, and then a two-force constant model is considered [Alfinito et al. (2005)].

This two-force constant model is introduced to describe the different flexibilities of the atomic bonds within the helices and within the loops of a transmembrane protein. These harmonic oscillations lead to an impedance noise, which is calculated as a function of temperature. The implications of this impedance noise on the ligand detection process are finally discussed.

### 4.2.1 *Classical Harmonic Oscillator*

To introduce the electrical fluctuations associated with thermal vibrations, two models are considered within the impedance network. Both of them use the probability distribution given by the classical harmonic oscillator.

In the first model, hereafter called the link oscillation model (LOM), the variation of the network characteristics (e.g. number of links, impedance, etc.) is related to the change in the distance between each pair of nodes.

In the second model, hereafter called the node oscillation model (NOM), the variation of the network characteristics is related to the isotropic variation of the position of each node in the three-dimensional space.

The block diagram of the algorithms used for the above two models is detailed in the Appendix.

### 4.2.2 *Link Oscillation Model*

In this model, the state of the protein is fully described by the matrix $D_0$, which contains the distances between each pair of nodes. To calculate the impedance of the structure, each distance is compared with the interaction length and, if it is less than $2R_a$, then the link is set with the impedance given by the corresponding elementary impedance model; otherwise, the link is cancelled. To describe the fluctuations, a number of random matrices $D$ are produced on the basis of the static matrix $D_0$ by making the distances fluctuating as a harmonic oscillator. The state of a harmonic oscillator at time $t$ is expressed as:

$$x = x_{max} \sin(\omega t) \tag{4.11}$$

where $x_{max}$ is the maximum amplitude and $\omega$ is the angular frequency of the oscillator.

The probability that $x_{max}$ takes the value $x_a$ at randomly selected times is proportional to:

$$P(x_a) \sim \frac{\partial t}{\partial x}|_{x=x_a} \tag{4.12}$$

The correspondence between a random number $r$ evenly distributed between 0 and 1 and $x_a$ is given by:

$$r = \frac{\int_{-x_{max}}^{x_{max}} P(x)dx}{\int_{-x_{max}}^{x_a} P(x)dx} \tag{4.13}$$

Accordingly, $x_a(r) = -x_{max} oos(\pi r)$ is obtained, or equivalently, since $r$ is uniformly distributed between 0 and 1, by:

$$x_a(r) = x_{max} \cos(\pi r) \tag{4.14}$$

Equation (4.14) can be used to randomly select the state of the harmonic oscillator. For the details, see the block diagram in Fig. A.3 of Appendix.

From the above, it follows that at each step the length of a link, $l$, is given by:

$$l = |l_0 + x_a(r)| = |l_0 + x_{max} \cos(\pi r)| \tag{4.15}$$

To consider a thermal fluctuation, according to classical statistics, the energy of the oscillator is taken to be proportional to $k_B T / x_{max}^2 \sim T$, and thus one takes:

$$x_{max} = A\sqrt{T} \tag{4.16}$$

where $A$ is a fitting parameter that, in general, depends on the elastic properties of the whole structure and thus on the distance between amino acids. As a first approximation, $A$ is taken to be the same for each pair of amino acids.

By recalling that in the impedance network model the impedance (absolute value) decreases at increasing values of $R_a$, one can construct the minimum impedance structure with characteristic radius of interaction $R_a + x_{max}$ and, correspondingly, the maximum impedance structure with radius $R_a - x_{max}$. A real structure will contain all links that satisfy the condition $l_0 < (R_a - x_{max})$, no links that satisfy the condition $l_0 > (R_a + x_{max})$, and some random links that satisfy the condition $(R_a - x_{max}) < l_0 < (R_a + x_{max})$. This approach preserves the shape of the structure and at the same time includes a source of fluctuation in the length of each link, which should be now regarded as an effective length appropriate to modulate the value of the corresponding impedance from Eq. (4.2).

### 4.2.3 *Node Oscillation Model*

In this model, nodes are allowed to fluctuate around their static position in an isotropic random direction. Accordingly, each random state is described by the array of the absolute coordinates of nodes $X, Y, Z$. For each $i$-th node:

$$X_i = X_0 + \delta x,\ Y = Y_0 + \delta y,\ Z = Z_0 + \delta z$$

$$\delta x = r \sin\theta \cos\phi,\ \delta y = r \sin\theta \sin\phi,\ \delta z = r \cos\theta$$

$$r = r_m \cos(\pi r_3);\ \phi = 2\pi r_2; \cos\theta = 2r_1 - 1$$

Here $X_0$, $Y_0$, $Z_0$ are matrices describing the static state, and $r_1$, $r_2$, $r_3$ are random numbers evenly distributed between 0 and 1. Then, on the basis of $R_a$ and this new system of nodes, the matrix of distances is built and the protein impedance determined. All other steps of the simulation are carried out in the same way as for the link oscillation model. For details one can see the block diagram reported in Fig. A.4 of Appendix.

### 4.2.4 *Results on Average Quantities*

This section reports the results of the average quantities: fraction of links and modulus of the impedance as functions of the maximum amplitude of the oscillations for the two classical oscillator models of the previous sections.

For the LOM, the number of links is calculated analytically as follows. The probability for a link to exist is:

$$P_i = \frac{1}{\pi}(\arcsin c_2 - \arcsin c_1) \tag{4.17}$$

where $c_1 = min(1, max(\frac{-2R_a - l_0}{x_{max}}, -1))$, and $c_2 = max(-1, min(\frac{2R_a l_0}{x_{max}}, 1))$.

Accordingly, the estimated average number of links in the fluctuating structure is equal to the sum of $P_l$ over all pairs of nodes. Figure 4.28, upper panel, reports the LOM results for BR at different values of the interacting radius. Theoretical values are found to be in perfect agreement with MC simulations. One should note that the number of links associated with a maximum amplitude systematically decreases on increasing the interacting radius.

For the NOM, the number of links is calculated only by MC simulations. Figure 4.28, lower panel, reports the NOM results for different values of the interacting radius.

One can see that contrary to the LOM, the number of links decreases with amplitude and increases with increase in the interacting radius.

The calculated average impedances versus $x_{max}/R_a$ are shown in Fig. 4.29 for the cases of a smooth and steep model of elemental impedances. Upper panel refers to LOM results and lower panel to NOM results, respectively. Here at increasing amplitude, a significant decrease in the impedance is found for the case of LOM in contrast to a significant increase in the impedance for the case of NOM.

### 4.2.5 *Variance of Impedance Fluctuations*

This section reports the results concerning the variance of impedance fluctuations as a function of the maximum amplitude of the oscillations for the smooth and steep models of the elementary impedance and for the LOM and NOM models. The calculated relative variance versus $x_{max}/R_a$ is shown in Fig. 4.30 for the LOM and for the smooth (upper panel) and the steep (lower panel) RC models, respectively. Analogously, Fig. 4.31 reports the results for the case of NOM.

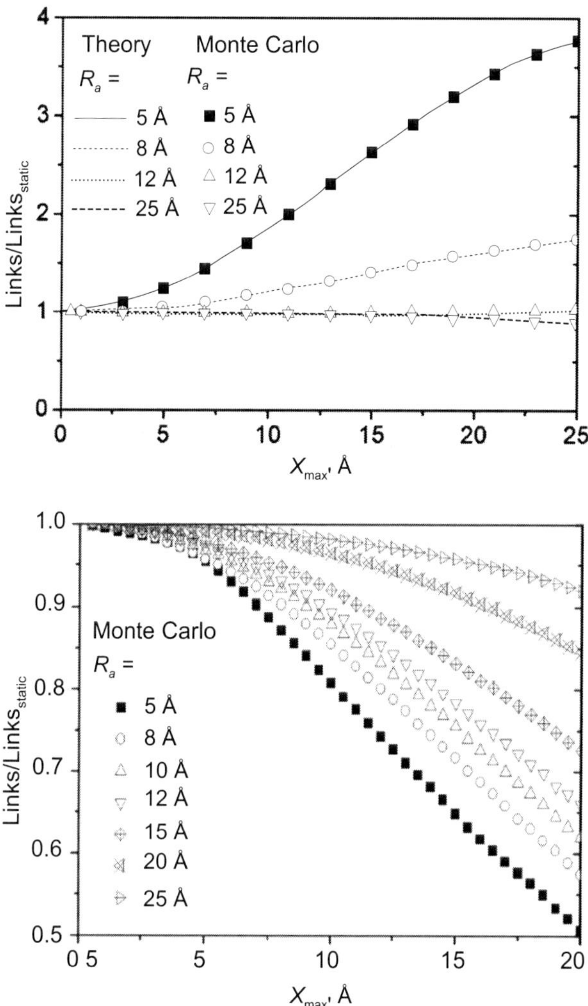


**Figure 4.28** Normalized average number of links versus the maximum amplitude for different $R_a$. Upper (lower) panel refers to the case of LOM (NOM). Continuous curves refer to analytical theory, and symbols refer to Monte Carlo simulations.

LINK OSCILLATION MODEL

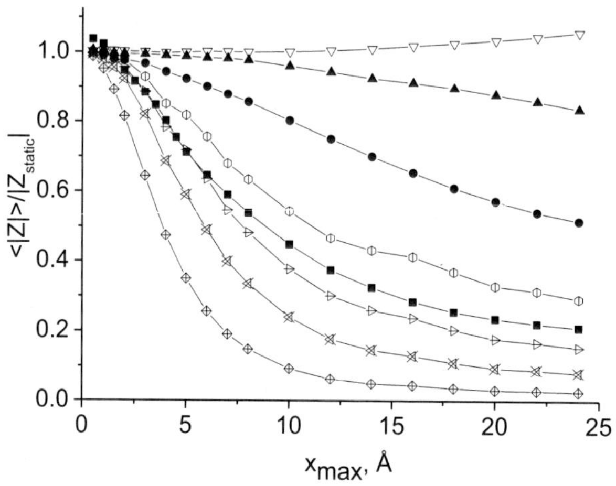


NODE OSCILLATION MODEL

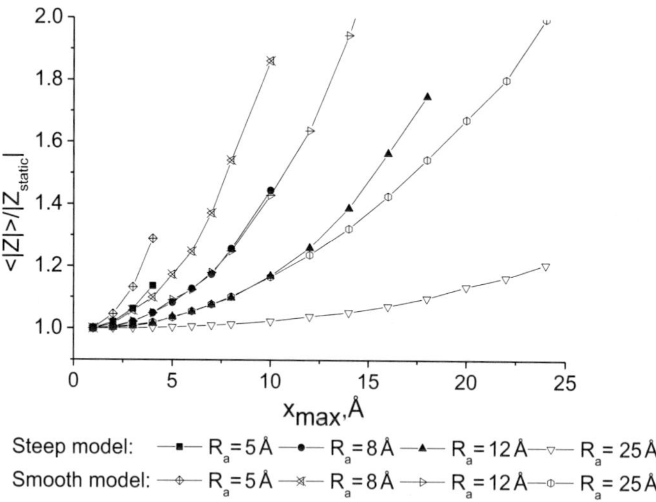


**Figure 4.29** Normalized impedance modulus versus $x_{max}$ for different $R_a$ and elementary impedance models. Upper (lower) panel refers to LOM (NOM). Symbols refer to Monte Carlo simulations, and lines are guides for the eyes.

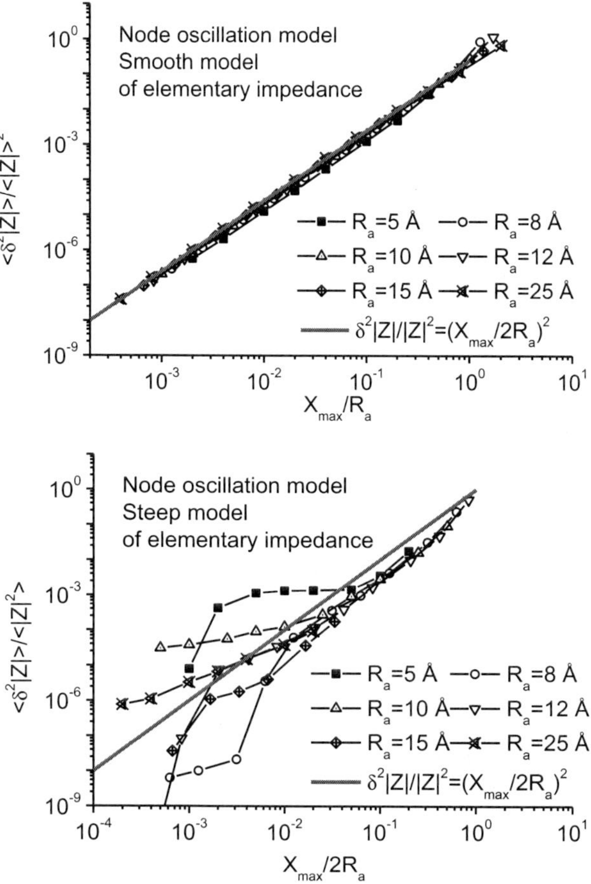


**Figure 4.30** Normalized variance of impedance-modulus fluctuations versus $x_{max}/R_a$ for the LOM. Upper (lower) panel refers to smooth (steep) RC model. Symbols refer to Monte Carlo simulations, and curves are guides for the eyes.

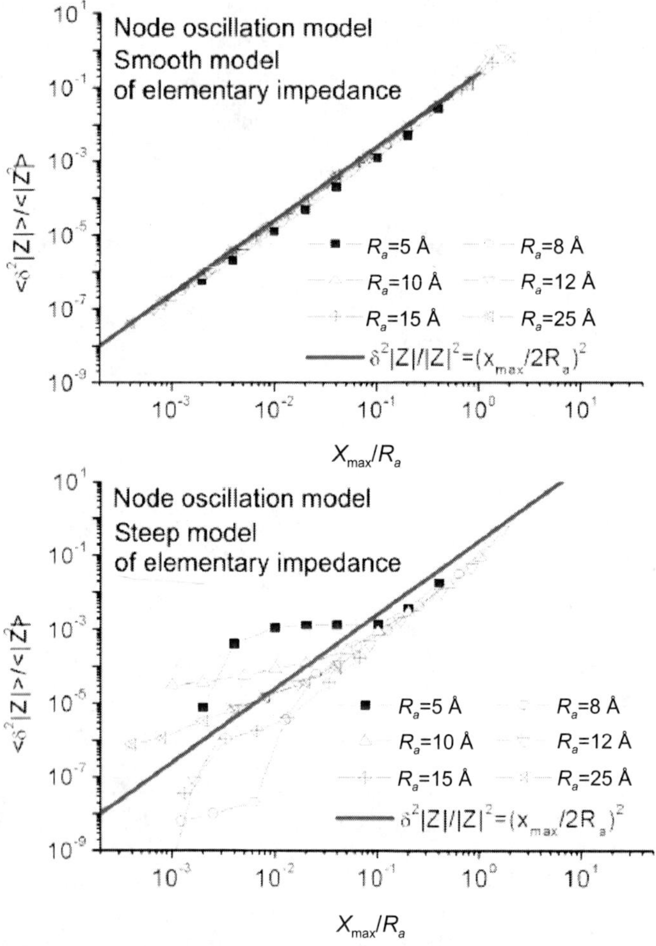


**Figure 4.31** Normalized variance of impedance-modulus fluctuations versus $x_{max}/R_a$ for the NOM. Upper (lower) panel refers to smooth (steep) RC model, and curves are guides for the eyes.

One should remind that $x_{max}$ can be transformed into a temperature within a multiplication factor. Furthermore, in the used algorithm, one should note that iterations are independent from each other and hence these graphs do not contain any time correlation but only amplitudes of statistical fluctuations. For the LOM it is found that the variation law for the smooth impedance model at small $x_{max}$ fits reasonably well with the expression:

$$\frac{\delta^2|Z|}{|Z^2|} = \left(\frac{x_{max}}{5R_a}\right)^2 \tag{4.18}$$

By contrast, the steep impedance model provides evidence of significant abrupt jumps, which are not predictable. One reason of these jumps is the presence of a telegraph noise due to the existence of bottlenecks in the link distribution inside the structure, which increases the variance at low $x_{max}$ depending on the structure peculiarities. Another reason is the weak sensitivity of the smooth impedance model to fluctuations, which dominates mainly at high values of $x_{max}$.

The steep impedance model provides results more sensitive to the structure peculiarities, and thus it should be of interest to discriminate among different structures of the same type. On the other hand, the smooth impedance model exhibits more general and uniform properties presumably related to the whole class of these structures.

For the NOM, the variation law for the smooth impedance model is found to fit reasonably well with the expression:

$$\frac{\delta^2|Z|}{|Z^2|} = \left(\frac{x_{max}}{2R_a}\right)^2 \tag{4.19}$$

It is the same expression as for the LOM but with a different numerical coefficient. This difference can be partly explained (within a factor of two) by the nonequivalence of the $x_{max}$ between the models. The remaining discrepancy is associated with a cumulative effect due to the influence of a node displacement on the displacements of all the connected distances. Again, the steep impedance model provides evidence of significant discontinuities that are a signature of the structure but are not predictable.

In conclusion, the results of simulations for the classical oscillator model can be summarized as follows: (i) The average quantities show an opposite behavior when considering the LOM or the NOM; (ii) an analogous behavior is found when considering the smooth and steep impedance models. The microscopic reason for such an opposite behavior remains at present an unsolved problem. The variance of impedance fluctuations shows (i) analogous behaviors for the LOM or the NOM; (ii) universal behavior for the smooth impedance model; and (iii) peculiar behaviors for the steep impedance model that means a signature of the network structure. The microscopic reason of the universal behavior remains at present an unsolved problem.

### 4.2.6 *Quantum Harmonic Oscillator*

This section reports the effect of thermal fluctuations on the electrical response to an AC bias within a quantum harmonic-oscillator approach. Accordingly, the nodes of the network are allowed to fluctuate around their equilibrium positions with an amplitude that depends on temperature. For the sake of simplicity and to get a first qualitative estimation, the system of fluctuating nodes is described as a set of independent, isotropic, harmonic oscillators, with common values of the force constant $\gamma$ and of the proper frequency, $\omega_0$, and in contact with a thermal bath at temperature $T$. By denoting with $\vec{r}_{n,eq}$ and $\delta\vec{r}_n = \vec{r}_n - \vec{r}_{n,eq}$, respectively, the equilibrium position and the displacement from the equilibrium of the $n$-th oscillator, the mean square displacement, $< (\delta\vec{r})^2 >$, is given by:

$$< (\delta\vec{r})^2 >= \frac{3}{2}\frac{k_B\theta}{\gamma} + \frac{k_B\theta}{\gamma}\frac{1}{\exp[\theta/T] - 1} \tag{4.20}$$

where $\theta = \hbar\omega_0/k_B$, with $\hbar$ the reduced Planck constant and $k_B$ the Boltzmann constant.

When $T \gg \theta$, Eq. (4.20) simplifies to:

$$< (\delta\vec{r})^2 > \approx k_B T/\gamma \tag{4.21}$$

For a given temperature, the wave function of the oscillator is a superposition of several excited states and its probability density cannot be expressed in a simple form. Again for simplicity, the probability density for the presence of the $n$-th oscillator around its equilibrium position is taken as:

$$P(\delta\vec{r}_n) = \frac{1}{[2\pi < (\delta x_n)^2 >]^{3/2}} \exp\left[-\frac{(\delta\vec{r}_n)^2}{2 < (\delta x_n)^2 >}\right] \tag{4.22}$$

where

$$< (\delta x_n)^2 >=< (\delta y_n)^2 >=< (\delta z_n)^2 >=< (\delta\vec{r})^2 > /3$$

According to the values commonly used in literature [Atilgan et al. (2001); Delarue (2002); Tirion (1996)], it is taken: $\gamma = 2.5$ KJ mole$^{-1}$ Å$^{-2}$, which implies $\theta = 12$ K. Thus, the condition $T \gg \theta$ is satisfied at room temperature. The choice of a unique force constant for all the amino acids of the protein is a common assumption in the literature, as for the Gaussian network models [Atilgan et al. (2001); Tirion (1996)]. On the other hand, it is well known that in the presence of thermal fluctuations, $\alpha$ helices are more stable than loops and terminals [Gether and Kobilka (1998); Lameh et al. (1990); Lefkowitz (2000)]. For this reason, by relaxing the above assumption, impedance fluctuations are analyzed by introducing two-force constants: $\gamma_1$ for $\alpha$ helices and $\gamma_2 < \gamma_1$ for loops and terminals. The values of $\gamma_1$ and $\gamma_2$ are chosen by fixing the mean value of the force constant to the previous value and then by considering different values of the ratio $\gamma_1/\gamma_2$. Therefore, the greater the value of this ratio, the more flexible the loops and terminals will appear with respect to the helices. The fluctuations of the impedance network are then calculated by an MC simulation that, starting from the static condition, at each step generates a random network with the position of nodes chosen stochastically according to a mean square displacement satisfying Eq. (4.20). The main results of this investigation are illustrated by the following figures. Figure 4.32 reports the modulus of the impedance normalized to its average value versus time for BR (upper panel) and metarhodopsin II (lower panel) at room temperature. The curves corresponding to two different values of the ratio $\gamma_1/\gamma_2$ are reported in grey (for $\gamma_1/\gamma_2 = 1$) and in black (for $\gamma_1/\gamma_2 = 20$). Figure 4.32 shows that the activated Meta II state is characterized by impedance fluctuations significantly wider than those of the Rho native state. This is particularly true when the force constant of the $\alpha$ helices is 20 times greater than the force constant of loop/terminals. This result reflects the minor thermal stability of the Meta II state with respect to the Rho state, probably due to an expansion of the entire structure, consequent to the conformational change [Menon et al. (2001)]. This conclusion is also confirmed by Fig. 4.33, which reports the probability distribution function $\phi$ of the impedance fluctuations at room temperature calculated for the native state (upper panel) and the activated state (lower panel), by

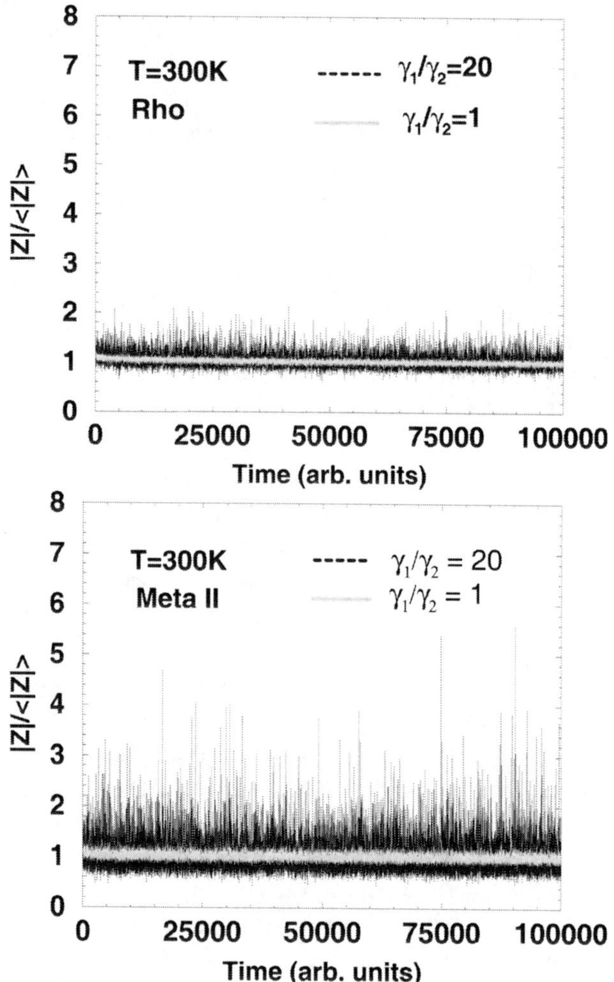


**Figure 4.32** Simulation of the impedance modulus versus time for BR (upper panel) and metarhodopsin II (lower panel). The impedance modulus is normalized to its average value, and the time is expressed in simulation steps. The grey and the black curves are calculated by taking the ratio $\gamma_1/\gamma_2$ equal to 1 and 20, respectively.

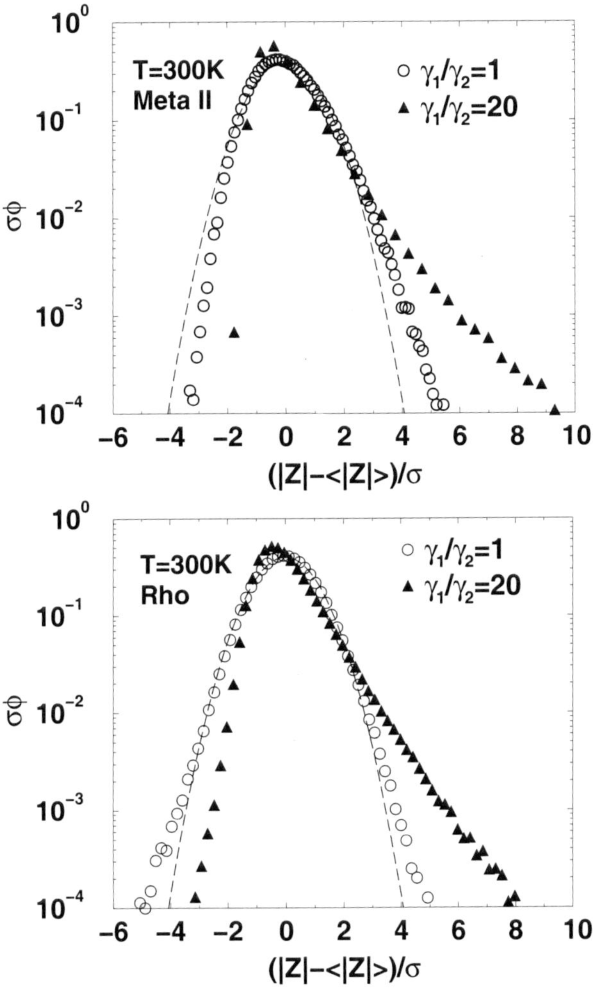


**Figure 4.33** Normalized PDF of impedance-modulus fluctuations at $T$ = 300 K for BR (upper panel) and metarhodopsin II (lower panel). The curves shown by circles and triangles are calculated by taking the ratio $\gamma_1/\gamma_2$ equal to 1 and 20, respectively. The dashed curve represents the Gaussian distribution.

taking the ratio $\gamma_1/\gamma_2$ equal to 1 (circles) and 20 (triangles). The probability densities are calculated by analyzing signals $Z(t)$ made of $5 \times 10^5$ data. Here $\sigma$ is the root mean square deviation from the average value of the impedance modulus $< |Z| >$. This normalized representation is adopted for reasons of convenience because it makes the distribution independent of its first and second moments [Pennetta (2004)].

Figure 4.33 shows that in the Meta II state, impedance fluctuations are non-Gaussian even when $\gamma_1/\gamma_2 = 1$ (for contrast, the fluctuations are Gaussian in the Rho state). Moreover, strong non-Gaussian tails are present when $\gamma_1/\gamma_2 = 20$ and in particular in the Meta II state. This strong non-Gaussianity at room temperature for $\gamma_1 \gg \gamma_2$ (and in particular in the Meta II state) is due to the presence of high spikes corresponding to the loss of links crucial to ensure the connectivity of the network.

Finally, Fig. 4.34 reports the calculated values of the relative root mean square fluctuation of the impedance modulus, $\sigma/ < |Z| >$, as a function of temperature. One can see that once the high flexibility of loops and terminals is accounted for by taking two different values for the force constants $\gamma_1$ and $\gamma_2$, the relative fluctuation of the impedance becomes strongly sensitive to the temperature. In particular, for $\gamma_1/\gamma_2 = 20$, the relative root mean square fluctuation of the impedance modulus in the Rho state is about 15%, while in the Meta II state, it is of about 35%. Such high levels of fluctuations would make problematic the detection of a conformational change in terms of variation of the impedance of a nanodevice, being these variations practically of the same order of magnitude [Pennetta (2004)]. However, the increase in the impedance noise itself in the Rho $\rightarrow$ Meta II transition (more than a factor of two at room temperature) should, in principle, provide a way to detect this transition.

From this investigation, the first conclusion is that a careful modeling of thermal molecular motion [Atilgan et al. (2001); Parak (2003)] is necessary to provide reliable estimates of the impedance noise whose level is crucial to the purpose of an electrical detection of the ligand capture by a GPCR. Furthermore, this simple model shows clearly that the transition Rho $\rightarrow$ Meta II is followed by a significant increase in the impedance noise level, which, in principle,

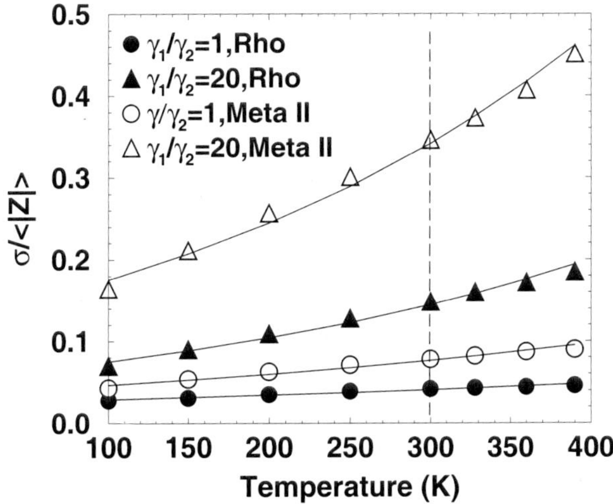


**Figure 4.34** Relative root mean square deviation of the impedance modulus as a function of the temperature for rhodopsin (full symbols) and metarhodopsin II (open symbols). The data shown by circles and triangles are calculated by taking the ratio $\gamma_1/\gamma_2$ equal to 1 and 20, respectively. The dashed vertical line provides evidence of the values of $\sigma/ < |Z| >$ at room temperature, and the continuous lines represent the best fit of the data with an exponential law.

would offer an interesting possibility to detect the transition itself using the logic that the noisier is the signal.

## 4.3 Current–Voltage Characteristics

This section reports an implementation on the theoretical model to investigate the nonlinear current–voltage (I–V) characteristics of a two-terminal sample where a given protein plays the role of the electrical active part. Accordingly, the I–V characteristics of a single protein are calculated by implementing the numerical code of the small-signal impedance and introducing sequential tunneling between neighboring amino acids as the microscopic mechanism responsible of charge transfer [Alfinito et al. (2011a); Alfinito and Reggiani (2013)]. The algorithm of the numerical code is detailed in the Appendix. The theory is then validated by comparing

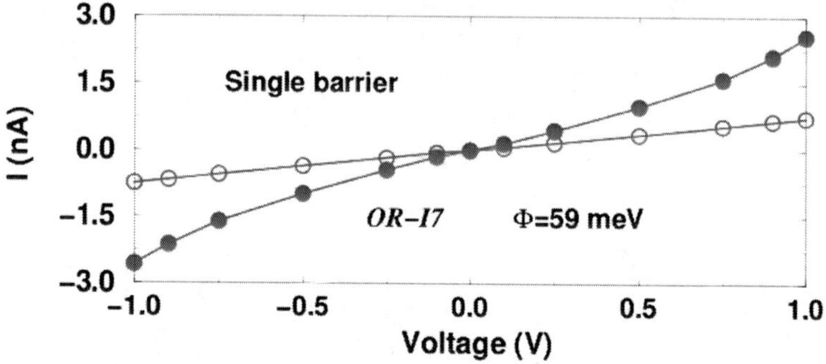


**Figure 4.35** I–V characteristics of the native and activated states of rat OR I7. Empty circles refer to the native state, and full circles to the activated state, respectively. The continuous lines are guides for eyes. Data are calculated with a single barrier height $\Phi = 59$ meV.

calculations with experiments available in the literature and carried out on different structures, as briefly illustrated in Chapter 3.

Figure 4.35 reports the I–V characteristics of rat OR I7, which are found to predict a substantial super-Ohmic behavior at increasing applied voltages better evidenced for the activated than for the native state of the protein. The parameters used for the simulation are the same used for bR $R_c = 6$ Å and an average barrier energy of 59 meV [Alfinito and Reggiani (2013)]. As found for bR, the activation of the protein is found to produce an enhancement of the current response.

# Chapter 5

# Bacteriorhodopsin as Testing Prototype

The nature of charge transfer (CT) in biological matter is a subject of outstanding interest both for fundamental knowledge and applications. Indeed, organic/biological devices are the new frontiers of technology, due to their potential characteristics of low cost, small size, and high specificity [BOND (2009–2011); SPOT-NOSED (2003–2005)]. In general, biological matter is not easy to be investigated, mostly because a standard (i.e., reliable and reproducible) way of preparation is still not available in most cases [Hou et al. (2006, 2007)]. Nevertheless, there are some relevant exceptions, such as monolayers of purple membrane (PM), a part of the cell membrane of the halophile *Halobacterium salinarum*, which is easy to be prepared and suitable for direct electrical measurements. PM is constituted by a single type of protein, the light receptor bacteriorhodopsin (bR), organized in trimers (see Fig. 1.10) and stabilized by lipids [Lozier et al. (1975); Luecke et al. (1999)]. For this reason, bR is chosen as a relevant prototype to investigate the correlation between its molecular structure and electrical properties as measured by different experimental techniques and theoretically interpreted within the impedance network protein analogue (INPA) model widely illustrated in Chapter 4.

*Proteotronics: Development of Protein-Based Electronics*
Eleonora Alfinito, Jeremy Pousset, and Lino Reggiani

ISBN 978-981-4613-63-7 (Hardcover), 978-981-4613-64-4 (eBook)
www.panstanford.com

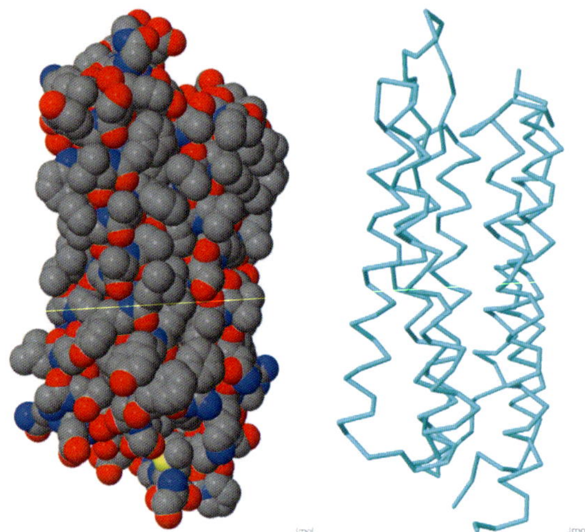

**Figure 5.1** Sphere (left panel) and backbone (right panel) representations of native bacteriorhodopsin 2NTU.

## 5.1 Modeling

Figures 5.1 and 5.2 report the sphere (left) and backbone (right) representation of bacteriorhodopsin in its native and active states as obtained by the homology modeling procedure.

## 5.2 Topological Properties

The protein topological properties are investigated by considering the native 2NTU and activated 2NTW structures of bR. The corresponding contact maps have been calculated for $R_c = 6$ Å and 12 Å, and the results are reported in Fig. 5.3 and Fig. 5.4, which show the respective contact maps.

## 5.3 Current–Voltage Characteristics

The entire structure of bR appears as a two-dimensional hexagonal crystal lattice, and its natural role is to use visible light for pumping

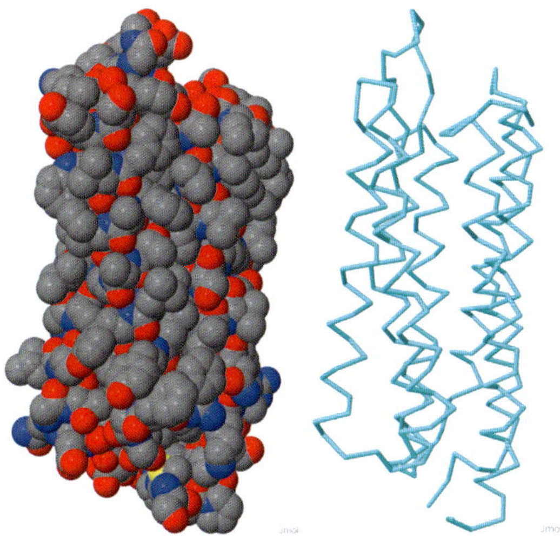

Figure 5.2 Sphere (left panel) and backbone (right panel) representations of active bacteriorhodopsin 2NTW.

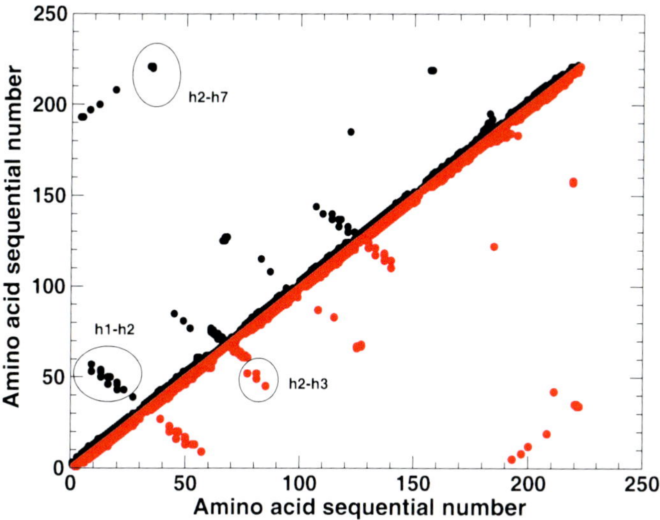


Figure 5.3 Contact map of the two bR representations: native state (2NTU on the left) and activated state (2NTW on the right). Because of the axial symmetry with respect to the diagonal exhibited by data, symbols pertaining to the native map are reported only on the left-hand side of the diagonal and those pertaining to the activated map on the right-hand side. The interaction radius is $R_c = 6$ Å.

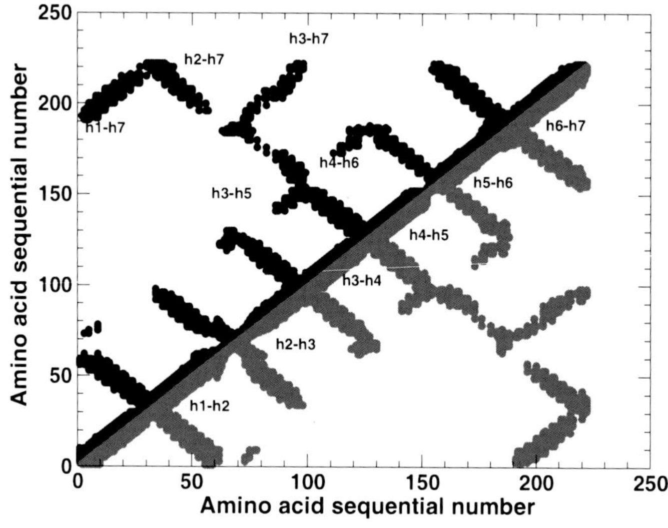


**Figure 5.4** Same as in Fig. 5.3 for an interaction radius of $R_c = 12$ Å.

protons outside the cell. In doing so, bR changes its tertiary structure (conformational change) and acts as a light-sensing protein. Current–voltage (I–V) characteristics of PM were analyzed under different experimental conditions [Casuso et al. (2007a,b); Jin et al. (2007, 2006)], giving clear evidence of super-Ohmic [Casuso et al. (2007a,b); Jin et al. (2007, 2006)] and illumination-dependent responses [Jin et al. (2007, 2006)]. These measurements are of relevant importance to understand the mechanism of CT, also in relation to protein activation. Furthermore, they are very promising for the development of a new generation of nanobiosensors that mimic the sensorial functioning of living species [BOND (2009–2011)].

The first experiment [Jin et al. (2006)] was carried on metal-insulator-metal (MIM) junctions of *millimetric* diameter, with the insulator constituted by patches of a 5 nm thick monolayer of PM. Measurements were carried out at bias up to about $\pm 1$ V with an observed maximum current of about 2 nA. The current response was found to be super-Ohmic and to increase up to about a factor of two when the sample was irradiated by green light. These results suggest that in this protein, like in some organic polymers [Zvyagin (2006)], CT is ruled by tunneling mechanisms. Furthermore, the strong dependence of the current from the presence/absence of the chromophore indicates the possibility of multiple carrier jumps across the protein, thus supporting the possibility for a CT transfer driven by sequential tunneling.

From a microscopic point of view, INPA describes the charge transport through a single protein by means of a *sequential-*tunneling mechanism between neighboring amino acids within an impedance network, which retains the main topological features of the protein, as reported in Chapter 4. In the present analysis, we chose $R_c = 6$ Å, a value that optimizes the native to activated state resolution, $\rho_{max} = 10^{20}\ \Omega$ Å, and $\rho_{min} = \times 10^{15}\ \Omega$ Å [Alfinito et al. (2011a)].

To model bR in its native state (in dark) [Alfinito et al. (2011a); Berman et al. (2000)], the Protein Data Bank (PDB) entry 2NTU is taken. The network is then connected to the external bias by means of perfectly conductive contacts, the input being taken on the first amino acid and the output on the last amino acid of the primary structure, respectively.

Figure 5.5 reports the comparison between experimental data [Jin et al. (2006)] and theoretical calculations carried out with two

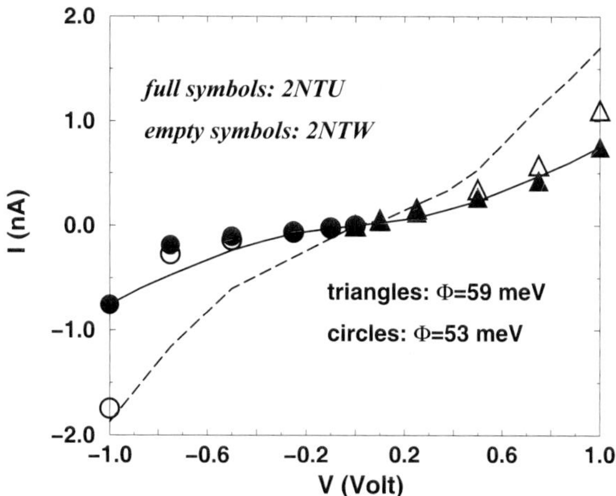


**Figure 5.5** I–V characteristic for bR in the native (2NTU) and activated (2NTW) states. Calculations are carried out with two different barrier heights, $\Phi = 59$ meV (triangles) and $\Phi = 53$ meV (circles). The continuous and dashed lines report the experimental data in the native and activated states, respectively.

different barrier heights, $\Phi = 59$ meV and $\Phi = 53$ meV, respectively. Because of the symmetry of simulated data with respect to the transformation V→ −V, the 59 meV data for positive bias and the 53 meV data for negative bias are reported. Despite the small difference in the barrier values, calculations exhibit significant difference in the current values at 1 V. In particular, the figure shows a larger resolution for the smaller $\Phi$ value, while the behavior of the native state is better reproduced by the larger $\Phi$ value. Furthermore, data reported in [Jin et al. (2006)] suggest large error bars on experimental curves (up to 50%), so that the estimation of the barrier height should be considered indicative [Alfinito et al. (2011a)].

Since in disordered organic materials charge transport can be described not due to a single but due to a Gaussian distribution of potential barriers [Bassler (1993); Zvyagin (2006)], the I–V characteristics were also calculated by using a Gaussian distribution of $\Phi$ values within an average value of 59 meV and two dispersion $\sigma$ values: $\sigma = 19$ meV and 94 meV, respectively. The results are reported in Fig. 5.6 where one can appreciate an effective improvement of the fit for the larger dispersion.

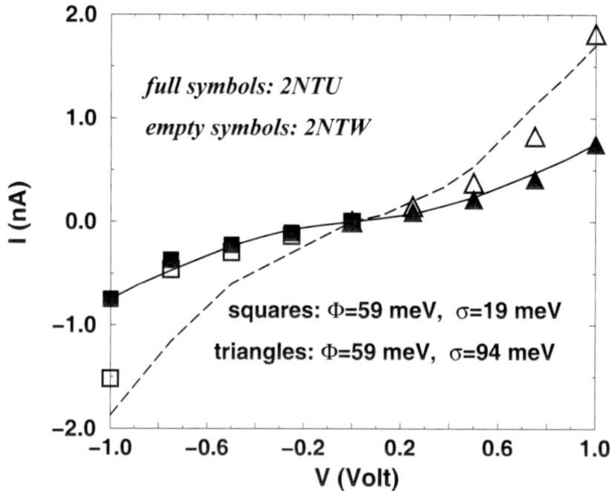


**Figure 5.6** I–V characteristic of bR in the native (2NTU) and activated (2NTW) states. Data were calculated by using a Gaussian distribution of barrier heights. The continuous and dashed lines report the experimental data in the native and activated states, respectively.

In the second experiment [Casuso et al. (2007a,b)], the I–V characterization was carried out at the *nanometric* scale, with the conductive atomic force microscopy (c-AFM) technique (see Chapter 3). Accordingly, one of the contacts was the tip of the c-AFM (100–200 nm of nominal radius). With respect to the first MIM experiment, in the common bias range, the current response was found to be lower for about four order of magnitude; thus the sample was able to sustain higher voltage up to about 5–10 V. Measurements were performed without any extra-light irradiation and with different tip indentations, starting from about 4.6 nm down to 1.2 nm. At voltages above 2 V, the presence of a cross-over between a direct tunneling (DT) regime and an injection, or Fowler–Nordheim (FN), tunneling regime [Simmons (1963)]. To model this second AFM experiment, the input simulates the extended tip of the AFM device. Accordingly, all the nodes with the $z$-coordinate (direction of the tip penetration), larger than that of the first amino acid, are given the same potential value. The output, point-like, remains on the last amino acid.

The values of $\rho_{max}$, $\rho_{min}$, and $\Phi$ are obtained by fitting the experiments of reference [Casuso et al. (2007a,b)] corresponding to an electrode distance of $L = 4.6$ nm. For this distance, the measurements are associated with a current crossing a single layer of proteins. Actually, it is used: $\rho_{max} = 8 \times 10^{13}\ \Omega$ Å, $\rho_{min} = 4 \times 10^{6}\ \Omega$ Å, and $\Phi = 219$ meV. The large difference between the $\rho_{max}$ and the $\rho_{min}$ values is dictated by the six order of magnitude spanned by the current values. The values of $\rho_{max}$ and $\rho_{min}$ imply the macroscopic resistance given by the $V/I$ ratio of the experiment. In such a view, the current measured at 1 V in reference [Jin et al. (2006)] yields a number of trimers equal to about $N = 10^9$ for a sample area of $2 \times 10^{11}$ nm$^2$, thus to a resistivity of about $10^{20}$ $\Omega$ Å for trimer [Alfinito et al. (2011a)]. By assuming, in present case, the same trimer resistivity, a number of trimers involved in the measured current of about $N = 10^6$ is estimated, thus leading to a crossing area of about $10^7$ nm$^2$. However, within an MIM electrical analogue, in reference [Casuso et al. (2007a,b)], the effective area deduced by the fit was found to be $0.1\times$ nm$^2$, about eight order of magnitude smaller than what estimated above. This dramatic difference should be mainly attributed to the MIM

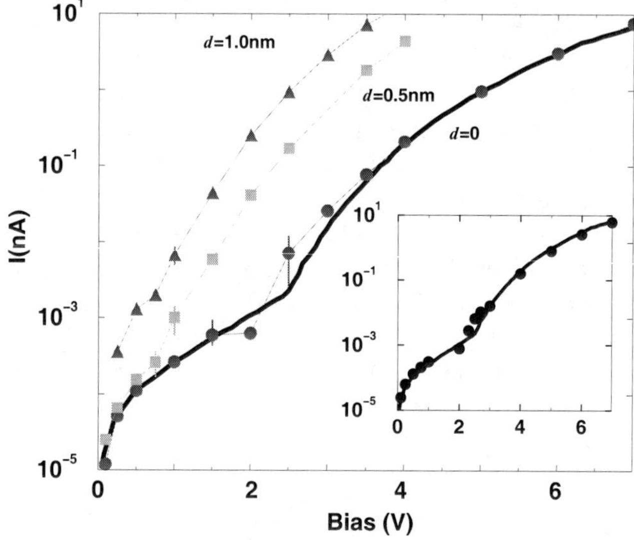


**Figure 5.7** I–V characteristics obtained by simulations with extended contacts at different indentations. The symbols and the thin continuous curves refer to calculations. The thick continuous line refers to the experimental data in the absence of indentation ($d = 0$) when the electrode distance is 4.6 nm [Casuso et al. (2007b)]. In the inset, the theoretical fit reported by symbols for $d = 0$ is performed with point-like contacts (see text).

electrical analogue used in [Casuso et al. (2007a,b)], which contrasts with the sequential-tunneling model used here.

Figure 5.7 compares the numerical and experimental data for the extended contact model and, in the inset, the point-like contact model at $L = 4.6$ nm. In both cases, the agreement is within the experimental and numerical uncertainty, and thus considered to be satisfactory. The barrier energy $\Phi = 219$ meV is taken to be independent of the contact choice.

When going from the point-like to the extended contact configuration, the fitting values of $\rho_{max}$ increase from $4 \times 10^{13}$ Ω Å to $8 \times 10^{13}$ Ω Å and also those of $\rho_{min}$ increase from $4 \times 10^{5}$ Ω Å to $4 \times 10^{6}$ Ω Å. Accordingly, the position of the extended contact is changed to reproduce the experiments obtained at different indentation of the tip.

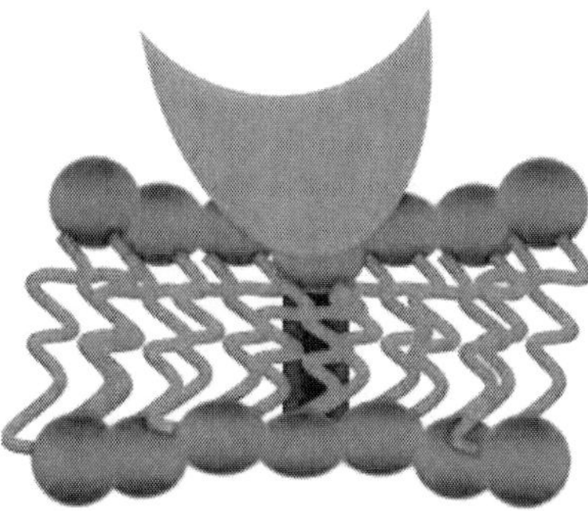

**Figure 5.8** Extended tip of the AFM device compressing the lipids and the protein.

Figure 5.8 reports a schematic of the tip indentation on a given protein. At increasing depths of the extended contact, the net effect is a reduced number of amino acids involved in the electrical transport. As a consequence, higher currents and a shift to lower potential values for cross-over between DT and FN tunneling regimes are expected.

Figure 5.9 reports the currents for different indentations when an Ohmic leakage current is added to the values reported in Fig. 5.9. The best fit is obtained by taking for the leakage resistance the values $8.4 \times 10^{12}$ Ω, $0.11 \times 10^{12}$ Ω, $0.036 \times 10^{12}$ Ω, respectively, for $L$ values of 4.6, 2.8, and 1.2 nm. The leakage resistance at $L = 4.6$ nm is taken equal to the protein resistance value at low bias. The decrease in the leakage resistance at increasing indentation can be related to geometrical effects associated with the decrease in the inter-electrode distance and the increasing surface of the tip contact when penetrating the protein. Interestingly enough, the microscopic interpretation is obtained with a length-independent electron effective mass, contrary to the case of reference [Casuso et al. (2007a,b)] where, to fit experiments, the carrier effective mass was increased over one order of magnitude on increasing the indentation. Furthermore, the value of the energy barrier is significantly reduced, for about a factor of 10, when compared with the values of the barrier height suggested in [Casuso et al. (2007a,b)]. Both these features are a consequence of the sequential-tunneling mechanism that replaces the single-tunneling

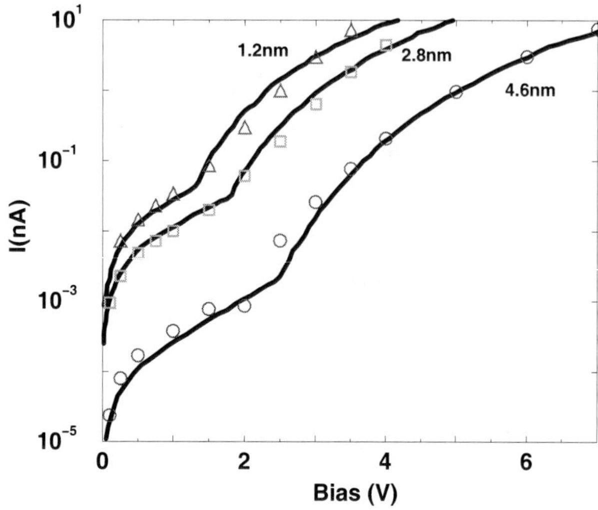


**Figure 5.9** I–V characteristics obtained by simulations with extended contacts at different indentations, including a leakage current. The thick continuous lines refer to experimental data [Casuso et al. (2007a,b)], and the symbols refer to numerical simulations.

mechanisms of the MIM model previously used [Casuso et al. (2007a,b)].

The INPA approach is also able to interpret the behavior of the I–V characteristics carried out with an *electrode–bilayer–electrode* structure and when the protein is illuminated or less by green light [Jin et al. (2006)]. To this purpose, INPA is applied to the PDB entry 2NTW (describing the activated state of bR) [Berman et al. (2000)] and calculations apply the same parameters used for the I–V characteristic of the native state in the extended contact configuration.

Figure 5.10 reports the simulated data calculated for both the native and activated states. The trend evidenced by experiments is reproduced here without introducing arbitrary parameters. Remarkably, present results are compatible with those reported in reference [Alfinito et al. (2011a)]. Here the higher value of $\Phi$ needed to fit the data of reference [Casuso et al. (2007a,b)] leads to some minor differences in the current responses of the native and

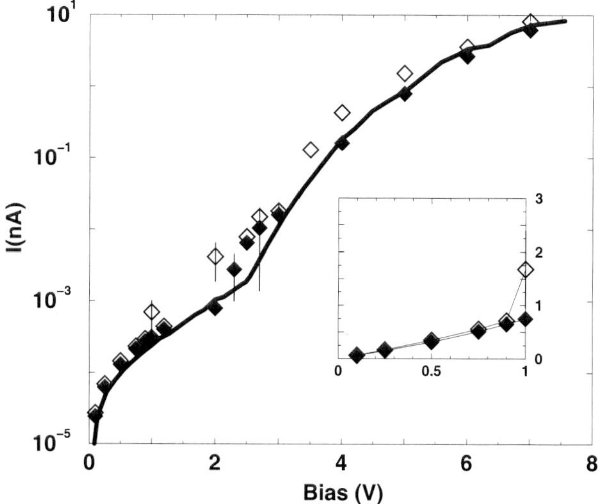


**Figure 5.10** I–V characteristics for the native and activated states of bR performed with point-like contacts. Closed and open symbols refer to native and activated states, respectively. Thick curve refers to experiments for the native state and for an electrode distance $L = 4.6$ nm [Casuso et al. (2007a,b)]. In the inset, the current is rescaled to the nA units to reproduce the value of the native state at 1 V measured in [Jin et al. (2006)].

activated states, which should be justified by the complexity of the physical system investigated.

In conclusion, the model allows for a consistent interpretation of a set of experiments carried out in a wide range of applied electrical potentials and in the presence or less of an external green light. The tertiary structure of the protein enters as direct data input and enables one to relate quantitatively the conformational change and the sensing action of the protein. Finally, also data obtained in high stressed conditions, like the penetration of an AFM tip in the protein membrane, can be finely reproduced. The qualitative and quantitative agreement between the numerical results and experiments poses the INPA, implemented for a sequential-tunneling mechanism, as a physical plausible model to investigate the electrical properties of other proteins, as will be reported in Chapter 6.

## 5.4 Scaling and Universality of High-Field Conductance in Bacteriorhodopsin Monolayers

Starting from the I–V characterization of the previous section, here the corresponding conductance and conductance fluctuations are investigated in a wide range of applied bias (up to about 20 V). To this purpose, the calculated instantaneous current is converted into the instantaneous chord conductance, $g = I/V$. Then the corresponding average conductance $< g >$ is compared with the experimental value inferred from I–V measurements, in the full range of applied voltages. Furthermore, conductance fluctuations around the average value, as obtained from numerical simulations, are analyzed in terms of their probability density functions (PDFs).

Interesting enough, when taking conductance as the order parameter, the analogous of a second-order phase transition is evidenced by the conductance versus voltage behavior near the crossing between the DT and the FN tunneling regimes. In particular, a critical value of the applied bias, say $V_C$, is identified and the reduced bias, $(V - V_C)$, is introduced as the new independent variable in place of $V$. Then it is observed that the conductance and the variance of its fluctuations increase and decrease with respect to the reduced bias following a power law with an exponent of about 3. The PDFs are found to be non-Gaussian and to pertain to the family of generalized Gumbel distributions, $G(a)$, which in the literature are associated with physical phenomena of very different nature [Alfinito et al. (2009c); Antal et al. (2009); Bertin (2005); Joubaud (2008); Noullez (2002)]. Accordingly, these results shed a new light on the microscopic mechanisms responsible for CT in transmembrane proteins and provide a novel physical example of a phase transition where the statistics is governed by the generalized Gumbel distribution [Bertin (2005); Clusel (2006); Noullez (2002)].

### 5.4.1 *Global Quantities*

To identify the macroscopic analogue of the critical voltage $V_C$ introduced in the previous section, from the set carrier frequency voltage-drop configurations compatible with the external applied

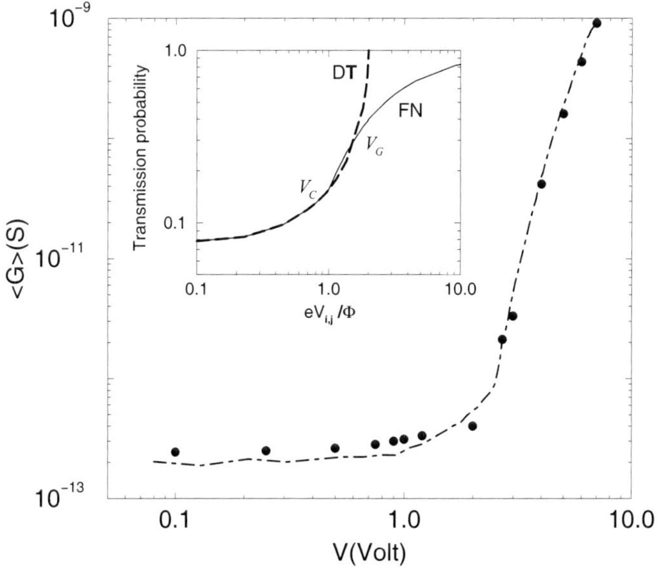


**Figure 5.11** Experimental (dot-dashed line) and calculated (full circles) data for the average conductance inferred from [Casuso et al. (2007a,b)]. Inset shows the transmission probability as given by DT (dashed line) and the interpolation of DT and FN (continuous line) for the typical parameters: $m_e$ the free electron mass, $l_{i,j} = 5.5$ Å, $\Phi = 219$ meV [Alfinito and Reggiani (2013)]. Here the values of the critical, $V_C$, and equivalent Ginzburg, $V_G$, voltages are indicated [Alfinito and Reggiani (2013); Alfinito and Vitiello (2002)].

bias, the average conductance, $< g >$, is calculated over a large ensemble of configurations and taken as the measurable global quantity.

Figure 5.11 reports the comparison between experiments and calculations where the model parameters used to get the best agreement are $R_C = 6$ Å, $\rho_{max} = 4 \times 10^{13}$ Ω Å, and $\rho_{min} = 4 \times 10^5$ Ω Å. After an initial region of quite constant value (Ohmic region), at about 0.7 $V$, $< g >$ starts to increase: Direct tunneling is here dominant and survives up to the equivalent Ginzburg voltage $V_G = 2.5$ $V$ [Alfinito and Reggiani (2013)]. In parallel, the FN tunneling starts to be relevant at about $V_C = 1.7$ $V$, then becoming the dominant mechanism at $V \geq V_G$.

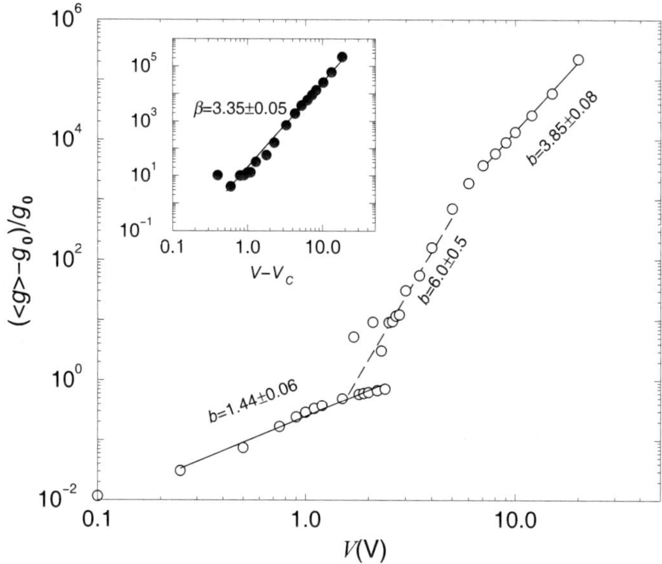


**Figure 5.12** Excess conductance from Fig. 5.11 versus applied voltage. Symbols refer to data from Monte Carlo (MC) simulations, and lines are power-law fits of data in different bias regions. The inset reports the excess conductance versus the reduced bias $V - V_C$, with $V_C = 1.7$ V.

Figure 5.12 reports the excess conductance $g_{EX} = (< g > - g_0)/g_0$, with $g_0 = 2.43 \times 10^{-13}$ $S$, the low field average conductance as a function of the applied voltage (open circles). The excess conductance $S$ found to increase as:

$$g_{EX} \propto V^b, \tag{5.1}$$

with the exponent $b$ being dependent on bias, as reported in Fig. 5.12. Remarkably, when $g_{EX}$ is shown as a function of the reduced bias, $(V - V_C)$, the high bias values of the exponent $b$ collapse in a single one, $\beta = 3.35$, as reported in the inset of Fig. 5.12. The initial dip seen in the inset signals the ending of an instability region just preceding the sharp increase in conductance. One should note that the value of the critical exponent $\beta = 3.35$ well compares with that of thermal excess conductance in superconductive thin films, in the regime of short-wave fluctuations (where $\beta = 3.2$) [Aswal et al.

(2002)], and resistance fluctuations in percolative systems (where $\beta = 3.7$) [Pennetta et al. (2002)].

Interesting enough, the normalized variance of conductance fluctuations, $S_g = (< g^2 > - < g >^2)/ < g >^2$, also gives evidence of a critical behavior in the range of voltages between $V_C$ and $V_G$, as reported in Fig. 5.13. In this region, a mixed regime (combined presence of DT and FN mechanisms) is found in which $S_g$ sharply increases over five orders of magnitude. This dramatic increase essentially determines the value of $< g >$, as seen in Fig. 5.11. Furthermore, the normalized variance of conductance fluctuations displays a behavior like:

$$S_g \propto V_c, \tag{5.2}$$

with the exponent $c$ being dependent on bias as reported in Fig. 5.13. After the rescaling $V \rightarrow (V - V_C)$, the intermediate and high bias exponents collapse into the single value $\gamma = -2.90$ that governs the power law of $S_g$, as reported in the inset of Fig. 5.13.

The above analysis leads one to the conjecture that all these scaling behaviors share a common origin, mainly due to the granular nature of the samples. As a matter of fact, in bR the amino acids represent the granular structure of the protein, similar to the case of superconductivity [Aswal et al. (2002)] where the exponent value is related to the grains inside YBCO (yttrium barium copper oxide) thin films, and to the percolative models where this behavior is associated with the formation of clusters [Stauffer and Aharony (1991)]. In support of the above conjecture, one should observe that the bR excess conductance and the variance of its fluctuations scale with practically the same exponent $\beta = -\gamma \sim 3$ at voltages above the critical value $V_C$, as shown in the insets of Fig. 5.12 and Fig. 5.13. In the framework of percolative models, a similar result agrees with the not-integer cluster dimension $d = 0.86$ [Campi (1986); D'Agostino (2003); Fisher (1967); Stauffer and Aharony (1991)] (Cantor dust).

One concludes that the voltage behaviors of conductance and of the variance of its fluctuations parallel those of a second-order phase transition typical of finite-size systems with an inertial region ($V_C - V_G$) [Bramwell (2009); Joubaud (2008)].

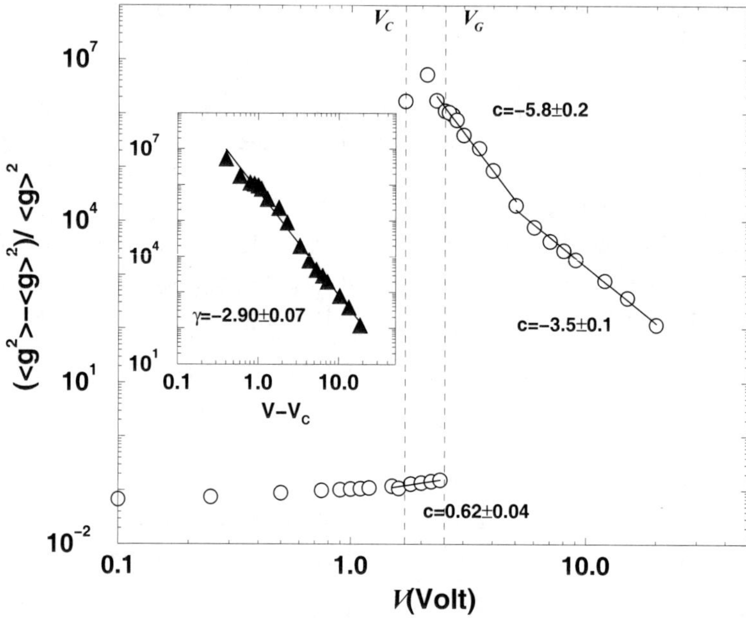


**Figure 5.13** Normalized variance of conductance fluctuations corresponding to the average conductance values of Fig. 5.11. Symbols refer to data from MC simulations, and continuous lines are fits with a power law of data in different bias regions. The inset reports the rescaled power law in the high bias regime.

## 5.4.2 *Generalized Gumbel Distributions*

To support and complement previous conclusions, the probability distribution functions of conductance fluctuations were calculated by following this procedure. The histograms of ln($g$) are collected for different bias values, in the range 0.1–9.0 V (see Fig. 5.14), and found to strongly deviate from a symmetric Gaussian-like shape, in the whole bias range. Furthermore, the non-Gaussian shape appears to be bimodal at the extremes of the considered bias range, while it becomes unimodal at intermediate and high bias values, i.e., close to the cross-over between DT and FN regimes and beyond. At the highest bias (i.e., above 9 V), the PDFs come back to the bimodal shape and, in this case, the two peaks appear well separated from each other.

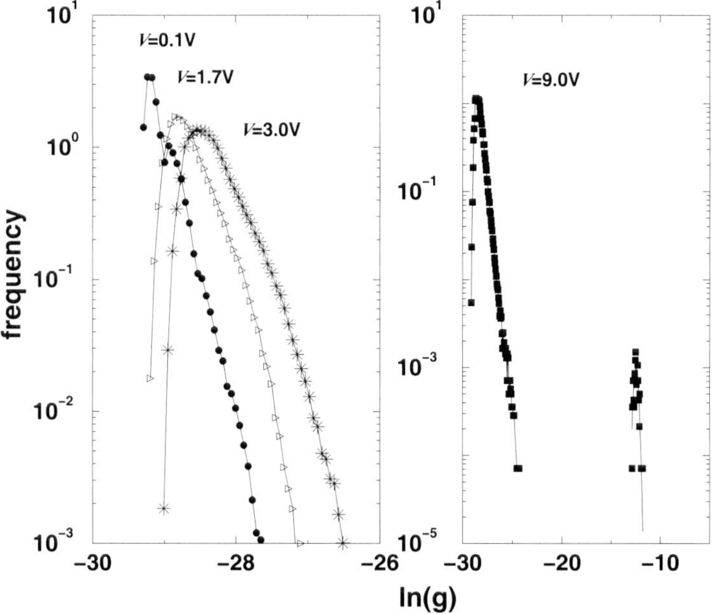


**Figure 5.14** Histograms of conductance fluctuations, in the range 0.1–9.0 V. Symbols refer to data from MC simulations, and continuous lines are guides to the eye.

By using a standard procedure of normalization [Bramwell (2009); Noullez (2002)], among the known ones, it is then possible to identify the best fitting PDFs. They are found to pertain to the class of generalized Gumbel distributions $G(a)$ [Bertin (2005); Clusel (2006); Noullez (2002)]:

$$G(a) = \frac{\theta(a)a^a}{\Gamma(a)} exp\{-aw - ae^{-w}\} \tag{5.3}$$

where $a$ is a positive numerical parameter, and $w$ is the real variable: $w = \theta(a)(z + \nu(a))$, with the scaled conductance $z = \frac{ln(g) - <ln(g)>}{\sigma_a}$, and $\sigma_a$ its root mean square value. The function $\nu(a)$ is defined in terms of the gamma function $\Gamma(a)$ and its derivatives as: $\nu(a) = \frac{1}{\theta(a)}(\ln(a) - \psi(a))$, with $\Gamma(a)$, $\psi(a)$, and $\theta^2(a)$ indicating, respectively, the gamma, digamma, and trigamma functions. Finally, $G(a)$ is a normalized distribution function with zero mean and unitary variance.

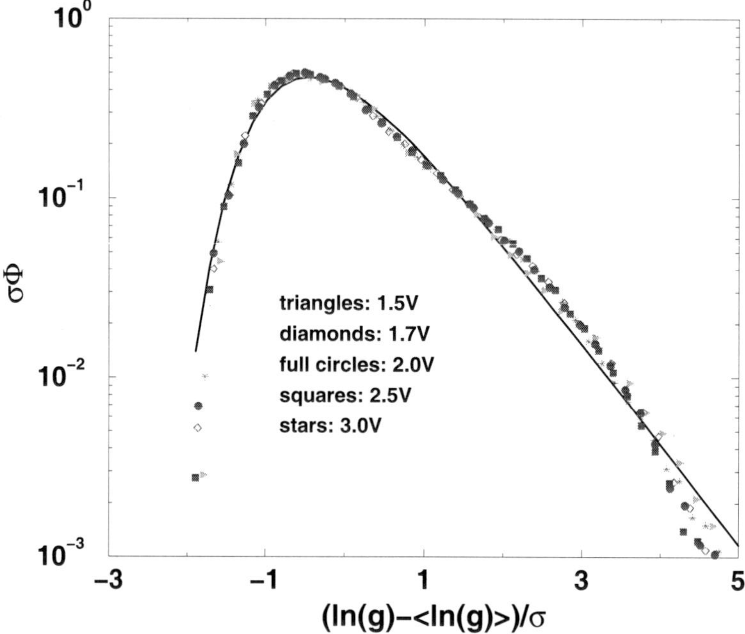


**Figure 5.15** Normalized PDFs in the range of bias values 1.5–3 V. Symbols refer to data from MC simulations, and the continuous curve is $G(1)$.

In the nucleation region, it is found that all the PDFs are well fitted by the "scaled Gumbel" $G(1)$ [Alfinito et al. (2012); Antal et al. (2009)] shown in Fig. 5.15. By further increasing the bias, this PDF smoothens down, converging toward the $G(0.7)$ function [Wolfram (2014)], as reported in Fig. 5.16.

The investigation of bimodal histograms requires a more careful analysis. The starting point is the ansatz that the corresponding PDF should be due to the superposition of two or more functions, all belonging to the $G(a)$ family, which in principle is not obvious [Bramwell (2009); Jaskierniak (2011)].

Then at low bias below 1 V, it is found that the bimodal histograms can be brought back to the $G(2)$ and $G(0.6)$ PDFs, as shown in Fig. 5.17. The bimodal shape is also present at very high bias, as shown in Fig. 5.17 and as can be inferred by the discussion

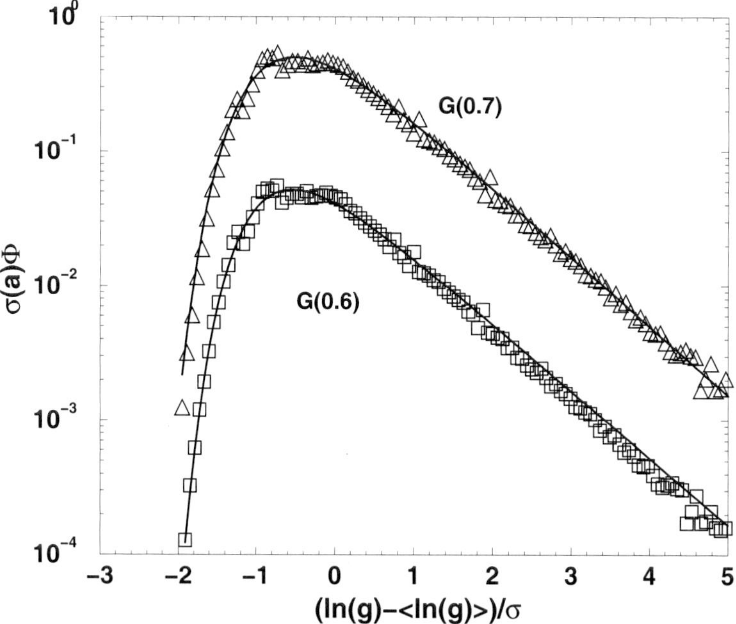


**Figure 5.16** PDFs of conductance fluctuations at 9 V. Symbols refer to data from MC simulations, and continuous curves report the $G(0.7)$ and $G(0.6)$ PDFs. To better resolve the differences among the distributions, the $G(0.6)$ curve and the related data are shifted by 0.1 on the vertical axis.

in the following section. Anyway, for the sake of simplicity, here only the low bias PDFs are reported.

### 5.4.3 *Discussion*

First, the question about the presence of skewed PDFs in the present system arises. The origin of this special feature should be searched in the presence of a low and a high boundary on the resistivity values. In particular, once the network is devised by choosing the protein structure and the value of the interaction radius $R_C$, the lowest conductance value is fixed. This value corresponds to the case in which all the resistivities coincide with $\rho_{max}$. In turn, this produces a negative skewness in the PDF, that is, the tail for positive fluctuation is much larger than that for negative fluctuations given

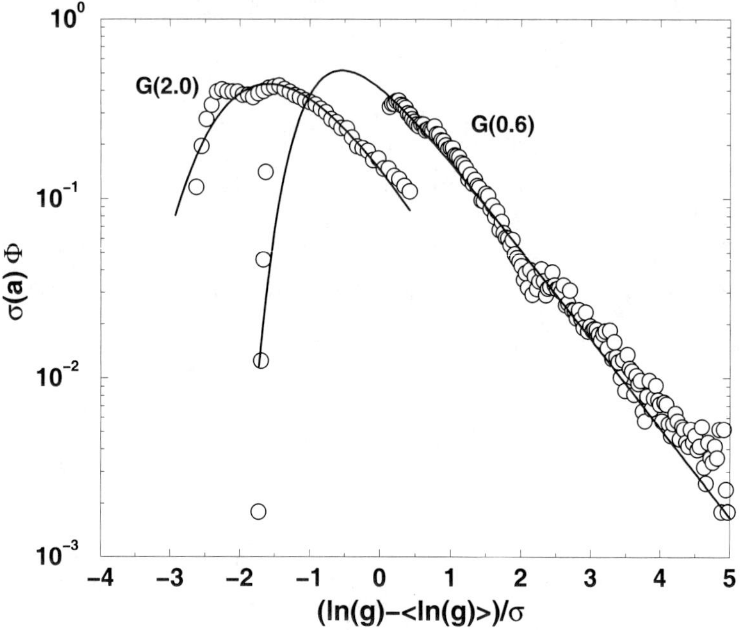


**Figure 5.17** PDFs at 0.1 V. Symbols refer to data from MC simulations, and continuous lines refer to the fitting functions $G(2)$ and $G(0.6)$.

by a symmetric distribution. By contrast, at a very high bias, the conductance reaches its maximum value, but it may also assume lower values: there is a positive skewness [Wolfram (2014)]. These constraints induce the presence of skewed PDFs all over the bias range. In particular, and due to their aforementioned origin, these PDFs belong to the family of extreme-event distributions.

Second, one should understand whether, and in the affirmative case how, the value of $a$ in the $G(a)$ PDFs is related to the system dynamics. To this purpose, it should be noticed that the generalized Gumbel distribution, Eq. (5.3), is equivalent, in the sense of distribution, to the gamma distribution of shaping parameter $a$ and lifetime $1/\lambda$ [Dufresne (2010)]:

$$f(t) = \frac{1}{\Gamma(a)}\lambda^a t^{a-1} e^{-\lambda t}. \tag{5.4}$$

The gamma distribution has been found relevant in many disciplines, going from economy to condensed-matter physics [Aste and Di Matteo (2008)] and is usually interpreted as the distribution of a sum of $a$ (if $a$ is an integer) identical and independent distributed (IID) variables with exponential distribution and the same lifetime. In other words, the sum of different gamma functions with the same lifetime is still a gamma function with a shaping parameter equal to the sum of the single shaping parameters [Akkouchi (2005)]. Otherwise, for IID variables with different lifetimes, the variable sum (convolution) can be expressed by means of gamma and beta distribution functions [Akkouchi (2005)]. In such a perspective, $G(2)$ should reflect the simultaneous presence of at least two main conductance paths among the multiple pathways constituting the resistance network [Alfinito and Reggiani (2013)]. In particular, the lowest conductance path corresponds to the quasi-insulating configuration of the network, where almost all the resistivities coincide with $\rho_{max}$. By contrast, a higher-conductance path is originated even at a very small bias by the low but finite probability for a network branch to take a low value of resistivity (see the inset in Fig. 5.11. Furthermore, a change in the resistivity value induces a variation in the distribution of local potential drops that propagate during the MC simulation. In this latter case, the mean conductance value follows from the stochastic evolution of the conduction paths [Landau and Binder (2009)].

Finally, the appearance of $G(2)$ reveals the simultaneous presence of two main conductance paths, with quite different origins (one mainly deterministic, the other mainly stochastic). This difference should be the reason of the not perfect superposition, i.e., not only $G(2)$ but also $G(0.6)$ is observed.

The less-than-one $a$ value of $G(0.6)$ could also be interpreted as describing a conduction path deprived of some components (low resistance branches). Indeed, the conventional interpretation of $G(a)$ with integer $a$ is the distribution of the $a$-th maximum/minimum of the system [Bertin (2005)]. As a matter of fact, in the crossover region, the $G(1)$ PDF signals the presence of a single dominant conductance path, due to the coexistence of multiple choices of resistivity values. This interpretation is also supported by the appearance of a $G(0.7)$ PDF in the region far from the crossover. In particular, $G(0.7)$ can be interpreted as a $G(1)$ deprived of some components belonging to the highest conductance values. The fluctuations around these highest values constitute the higher peak of the total PDF.

As a final remark, Eq. (5.3), for $t \to s^{\mu}$, takes the same form of the cluster distribution $n(s)$ [Fisher (1967); Stauffer and Aharony (1991)]:

$$n(s) \propto s^{-\tau}\, e^{-\lambda s^{\mu}}, \qquad \tau = (1 - \mu a). \tag{5.5}$$

This equation is widely used to describe critical-like systems from percolation to liquid-gas phase transition and to nuclei fragmentation [Campi (1986); D'Agostino (2003)].

### 5.4.4 *Conclusion*

The main results concerning conductance and its fluctuations can be summarized as follows. In a monolayer of bacteriorhodopsin, the voltage-dependent conductance is found to exhibit a second-order phase transition with a nucleation/inertial regime. This critical behavior is associated with the cross-over from a charge transport controlled by a DT mechanism to that controlled by a FN tunneling mechanism. The phase transition is confirmed by the existence of power laws that fit the conductance and its fluctuations as functions of the applied voltage and by a predicted sharp increase (over five order of magnitude) in the variance of conductance fluctuations. The PDFs of conductance fluctuations are found to follow the shape of a generalized Gumbel distribution $G(a)$ (see Eq. (5.3)). These distributions are non-Gaussian, exhibit both unimodal and bimodal shapes, and show a universal feature by an appropriate scaling of the variables. The bimodal PDFs can be decomposed into at least two unimodal distributions, and in general, the shaping parameter $a$ is a function of the applied bias. Despite being the first step, it is believed that the electrical properties of bR analyzed here open a scenario of relevant interest for other biomaterials besides those belonging to the family of transmembrane proteins.

# Chapter 6

# Survey of Other Proteins

The unified impedance network protein analogous (INPA) modeling described in Chapter 4 is applied here to a set of proteins that received particular attention in the literature and for most of which experimental data on their electrical properties are available. Accordingly, the investigation is carried out using both macroscopic approaches (Langmuir equation, etc.) and the microscopic INPA model already detailed in Chapter 4.

## 6.1 Proteorhodopsin

This section investigates the static and dynamic electrical properties of proteorhodopsin (pR) in connection with its 3D structure and provides a comparative analysis with its companion bacteriorhodopsin (bR). Proteorhodopsin is the most widely diffused retinal-based protein and seems to have a relevant role in preserving the correct equilibrium in marine ecosystem [Beja et al. (2001); DeLong (2010); Dioumaev et al. (2002)]. It was originally found in an uncultivated marine bacterium of the *SAR86* phylogenetic group [Beja et al. (2001)], and it is quite clear that mutants of this protein are worldwide present in bacterioplankton [Beja

*Proteotronics: Development of Protein-Based Electronics*
Eleonora Alfinito, Jeremy Pousset, and Lino Reggiani

ISBN 978-981-4613-63-7 (Hardcover), 978-981-4613-64-4 (eBook)
www.panstanford.com

et al. (2001); Dioumaev et al. (2002); Sabehi (2004)], but also in Archea and eukaryotic marine protists [Bamann et al. (2013)]. Furthermore, it has been observed that different kinds of pRs found in marine species can adapt to different wavelengths, at different ocean depths [Beja et al. (2001); DeLong (2010); Riesenfeld (2004)]. It is also remarkable that the present technology is able to produce protein mutants by heterologous expression in cells like *Escherichia coli*, and that these proteins are able to react to different wavelengths and also to be engineered for producing light flashes [Kralj et al. (2011)]. Regarding the mechanism of proton pumping, consequent to light activation, proton transport has been observed, depending on the value of the environmental pH concentration, both inward and outward the cell membrane [Lörinczi et al. (2009)]. Moreover, experiments performed in electrochemical chambers containing bacteria in which pR was expressed revealed a significant improvement in current generation during illumination [Johnson et al. (2010)].

Light-activated proteins such as pR and bR are conjugated with a retinal molecule, located among the seven transmembrane helices. The retinal isomerizes going from the *all-trans* to the *13-cis* configuration when photons of the visible spectrum hit on it. As a consequence of this change, a cycle of transformations involves the protein as a whole, driving it from the native to the activated state and, in turn, back to the native state. During this cycle, a proton is first released across the membrane out of cell and then recollected. The structural change is expected to produce some modifications in the electrical properties of the protein; this aspect is crucial for any possible use of this (or similar) protein in light-converter devices [Alfinito et al. (2009c, 2011a); Alfinito and Reggiani (2009b, 2013)].

To explore these features, some studies were initially performed on bR, as reported in Chapter 5. Later on, measurement of I–V characteristic in pR was described [Melikyan et al. (2011)]. The experiment was performed on a thin film device with a contact distance of about 50 μm, that is, $10^4$ times longer than that used in bR [Casuso et al. (2007a); Jin et al. (2006)]. Therefore, a direct comparison between data obtained in pR and in bR is not easy to be carried out. Anyway, in both these cases, a significant result

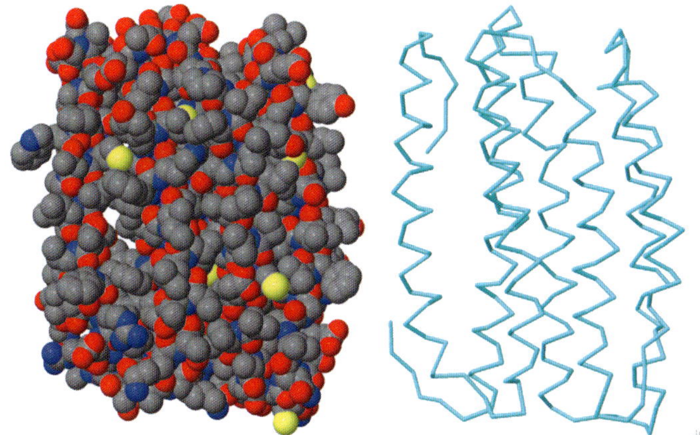


**Figure 6.1** Sphere (left panel a) and backbone (right panel b) representations of native proteorhodopsin 2L6X.

of a photocurrent was confirmed and found of great interest for optoelectronic applications.

Below, the electrical response of pR in a metal-insulator-metal (MIM) device is reported, and then the obtained data are compared with those of bR.

### 6.1.1 *Modeling*

Figure 6.1 reports the sphere (left) and backbone (right panel b) representation of pR in its native state as obtained by the homology modeling procedure.

### 6.1.2 *Topological Properties*

In the present case, the only interaction that was taken into account is that associated with the transfer of charge, and $R_c$ is taken in the range of values 5–15 Å. Accordingly, it was shown [Alfinito et al. (2008, 2011a); Alfinito and Reggiani (2009b)] that the best agreement between structure and function is obtained for $R_c$ values within the range 5–40 Å. On the side of the protein structures, while more and more refined models were proposed both for the native and activated states of bR, at present there is only one certified set

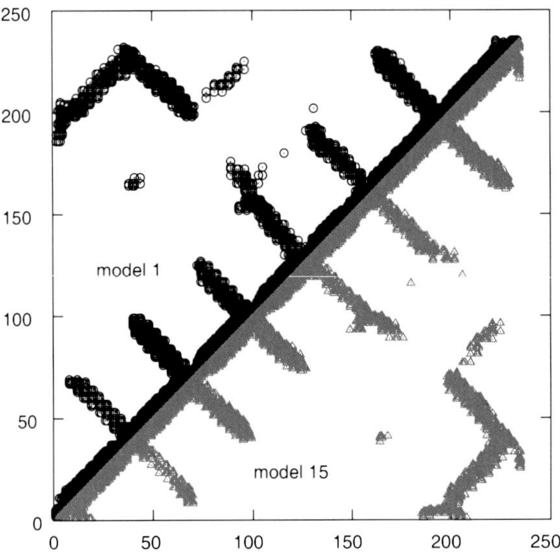


**Figure 6.2** Contact maps of model numbers 1 (on the left) and 15 (on the right) of a single protein of pR in its native state taken for $R_c = 12$ Å.

of structure models of pR in its native state. This set, taken with a nuclear magnetic resonance (NMR) technique and corresponding to the 2L6X entry of the Protein Data Bank (PDB) [Berman et al. (2000); Reckel et al. (2011)], consists of 20 different models, all in principle equally able to describe the protein electrical properties.

In the following, protein modifications are investigated by means of different tools, such as analysis of the contact maps, determination of the link distributions, and estimation of the protein global resistance. The differences of structure in the 20 models of PDB entry 2L6X are analyzed by means of the contact maps, which, for an assigned $R_c$ value, show the connected amino acids. For the sake of simplicity, only the analysis performed on a few of the 20 models, taken as exemplary cases, is reported here.

Figure 6.2 shows the contact maps of models 1 and 15 of pR traced for $R_c = 12$ Å. Even if significant differences can be observed between models 1 and 15, in particular, in the central region of the protein (the same is found for other models, not shown here), in the following, model 1 is taken as the reference

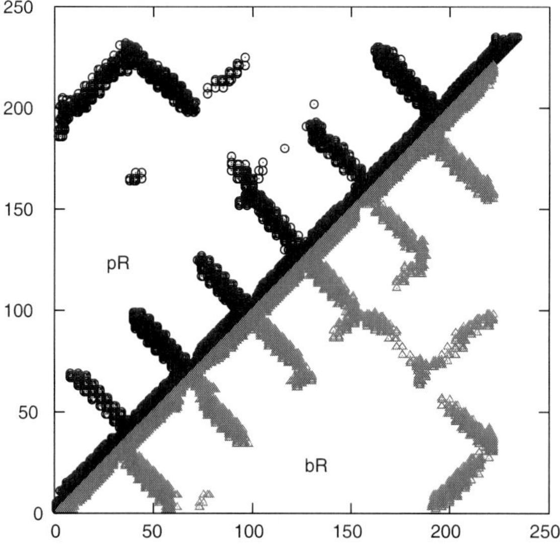


**Figure 6.3** Contact maps of model number 1 of a single protein of pR (on the left) and bR (on the right) in their native state taken for $R_c = 12$ Å.

structure. When comparing pR with bR, as reported in Fig. 6.3, the similarities between these proteins clearly emerge and the remaining differences are found to be more significant near the central zone of the protein.

Further information can be gained by checking the degree distribution, that is, by counting the number of links for each amino acid, at different $R_c$ values. Results are reported in Fig. 6.4 and Fig. 6.5. In particular, Fig. 6.4 compares the degree distributions of three different pR models for two values of $R_c$. For all the analyzed models, the maximum degree is 6 at $R_c = 6$ Å. In other terms, most amino acids, independent of their position in the protein, have the same number (6) of nearest neighboring, an unexpected regularity. Furthermore, the distribution rapidly smoothens down at increasing $R_c$ values.

Figure 6.5 reports the comparison between the degree distributions of model 1 of pR and bR. As already noticed for the contact maps, in this case also the structures appear quite similar, although some relevant differences stand out.

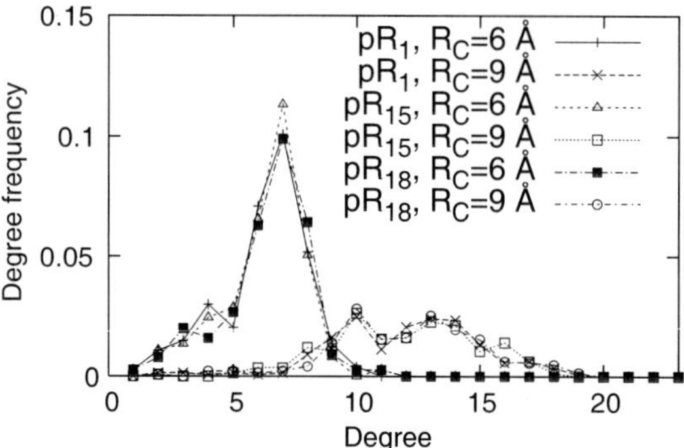


**Figure 6.4** Degree distributions for the native state of different pR models.

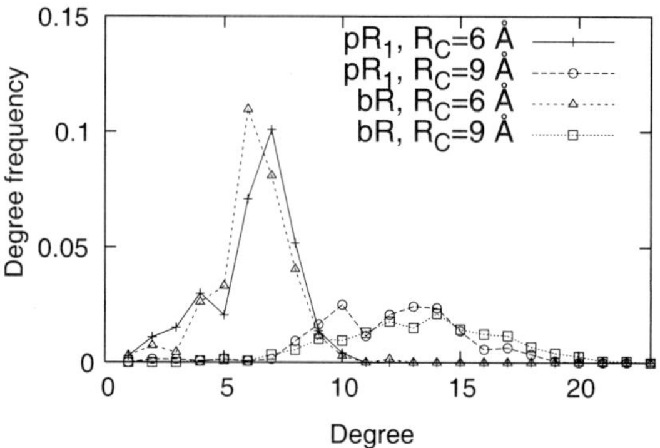


**Figure 6.5** Degree distribution for the native state of pR (model 1) and bR, for different $R_c$ values.

### 6.1.3 *Experiments*

In the wake of the breakthrough results on bR monolayers (width of about 5 nm) [Casuso et al. (2007a,b); Jin et al. (2006)], an interesting set of measurements have been performed on a MIM thin film of pR [Melikyan et al. (2011)]. In this latter experiment, the channel width of the film active region is of about 50 μm, sandwiched between two gold electrodes evaporated on a glass substrate with thickness of about 200 nm and length of about 3 mm. Accordingly, the buffer solution and the amount of pR, ranging from 0 to a maximum of 1 OD (optical density), occupy an active volume of about $3 \times 10^{-8}$ cm$^3$.

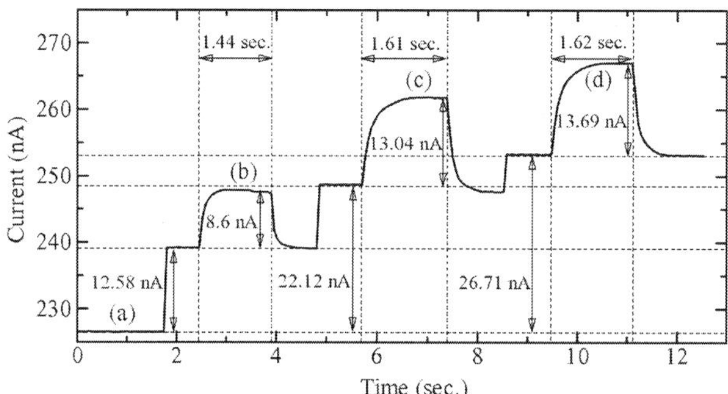


**Figure 6.6** Real-time measured photocurrent for (a) buffer alone; (b)-(d) buffer and pR with concentration of 1 OD, 5 OD, and 10 OD under 150 $mW/cm^2$ incident light intensity at 625 nm wavelength. The voltage applied to the MIM structure was 120 V. The vertical dashed lines are marks to show the exposure times in every case.

Figure 6.6 shows the real-time measured photocurrent for four samples with different pR concentrations, respectively, of (a) buffer alone and buffer plus pR with concentrations of (b) 1 OD, (c) 5 OD, and (d) 10 OD for 150 $mW/cm^2$ incident light intensity at 625 nm wavelength and 120 V fixed voltage. The observed current level is constant (227 nA) for buffer alone but increases when adding pR with different concentrations and also when, for a given pR concentration, illuminated with an incident light. The current increases over the buffer level in dark conditions are, respectively, 12.58 nA, 22.12 nA, and 26.71 nA. From Fig. 6.6, the estimated photocurrents are found to be 8.6 nA, 13.04 nA, and 13.69 nA, respectively, for 1 OD, 5 OD, and 10 OD. One should also note that a higher photocurrent is associated with a higher exposure time, the exposure time being the interval during which the generated photocurrent increases from zero to the saturation level. For pR concentrations of 1 OD, 5 OD, and 10 OD, the exposure times are, respectively, 1.44 s, 1.61 s, and 1.62 s, that is, by increasing pR concentration, photocurrents and exposure times also increase, and a saturation effect sets in for concentrations higher than 5 OD.

Figure 6.7 reports the experiments performed to investigate the photocurrent response of 10 OD of pR at increasing light intensity, from 40 to 120 $mW/cm^2$, and 120 V applied voltage.

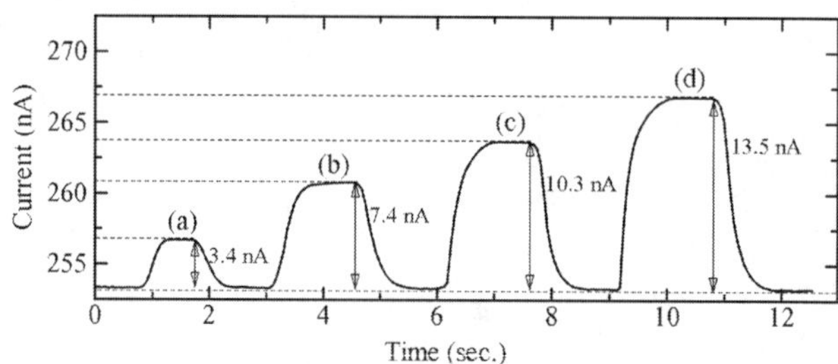


**Figure 6.7** Real-time measured photocurrent of 10 OD pR for incident light intensity of (a) 40 mW/cm$^2$, (b) 60 mW/cm$^2$, (c) 120 mW/cm$^2$, and (d) 150 mW/cm$^2$ at 625 nm wavelength and an applied voltage of 120 V.

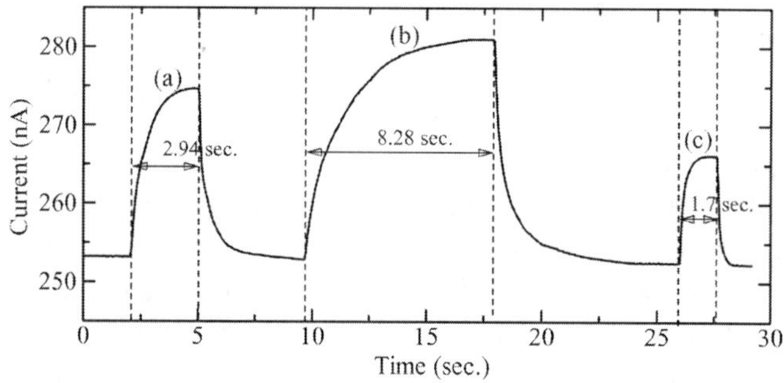


**Figure 6.8** Real-time measured photocurrent of 10 OD pR at incident light of wavelengths (a) 472 nm, (b) 532 nm, and (c) 625 nm. The incident light intensity and applied voltage are 150 mW/cm$^2$ and 120 V, respectively.

As a test of the pR sensitivity to different light wavelengths, Fig. 6.8 reports the time diagram of the output photocurrent of 10 OD pR at the incident light, respectively, of wavelengths (a) 472 nm, (b) 532 nm, and (c) 625 nm. The intensity of the incident light was 150 mW/cm$^2$ for an applied bias of 120 V. Here three independent but similar measurements are combined for comparison with each other. By changing the wavelength in the visible range, the exposure time of the generated photocurrent varies in correspondence with the variation of the absorption spectra of pR; the longer the exposure time, the higher the absorption coefficient. Accordingly, the photogeneration phenomenon is slowest in the green range (8.28 s), faster in the blue range (2.94 s), and finally the fastest in the red range (1.7 s), as shown in Fig. 6.8. A possible explanation of this behavior relies on a photoconductive behavior of samples where the gain-bandwidth product is constant: The photocurrent signal then becomes higher as the time response becomes longer. Furthermore, for the different wavelengths, the photocurrent amplitude is in agreement with the spectral dependence of the absorption coefficient reported in reference [Melikyan et al. (2011)].

The main results contained in the previous figures are briefly summarized as follows. Figure 6.6 proves the existence of a contribution to the dark current due to the presence of pR (in addition to that due to the buffer only) and also the existence of a net photocurrent in the presence of a visible light, still attributed to the presence of pR. Both these dark and light current contributions increase at increasing pR concentration. The latter outcome can be described within the Langmuir adsorption equation [Langmuir (1916)] that predicts a saturation after an initial linear increase in the current as:

$$I = G_{\text{tot}} V \tag{6.1}$$

with

$$G_{\text{tot}}(C) = G_{\text{sub}} + \frac{C}{1+C} G_{\text{pRsat}} \tag{6.2}$$

where $C$ is the pR concentration in units of OD.

Equation (6.1) is used in Fig. 6.9 to fit currents in dark (open circles) and in light (full circles) as a function of pR concentration.

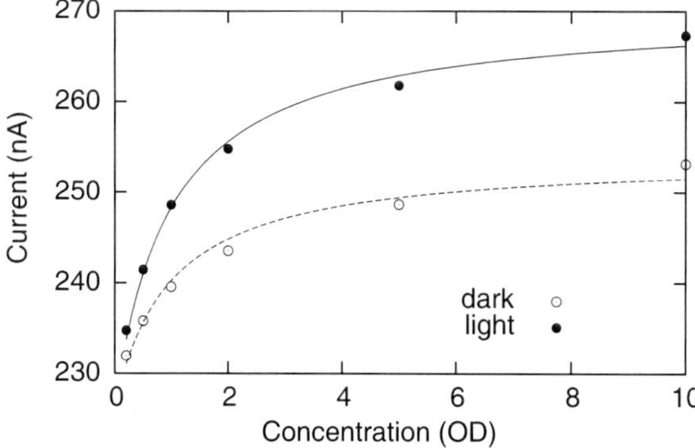


**Figure 6.9** Current as a function of pR concentration at 120 V applied voltage. Open (full) circles refer to dark (illuminated) conditions and dashed (continuous) curves to a fit of experiments obtained with the expression Eq. (6.3) reported in text.

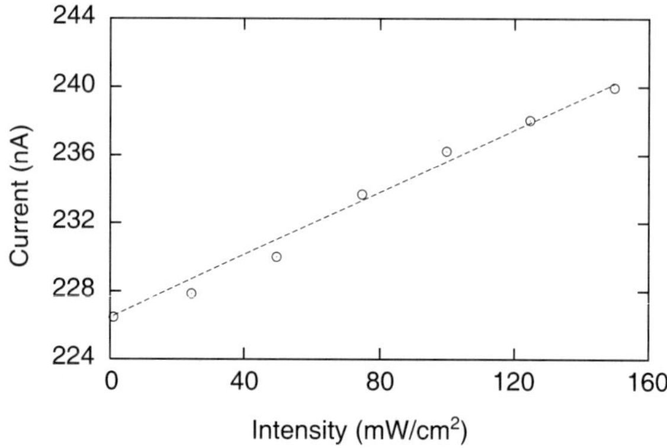


**Figure 6.10** Total current flowing through the device as a function of light intensity for an applied voltage of 120 V and a pR concentration of 10 OD.

In particular, it was chosen $G_{\mathrm{sub}} = 1.89$ nS and $G_{\mathrm{pRsat}} = 0.229$ nS for the dark and 0.363 nS for the illuminated conditions with an intensity of 150 mW/cm$^2$.

Figure 6.10 shows a quasi-linear increase in photocurrent at increasing light intensity. These data are described in terms of the following expression:

$$I_{\mathrm{tot}}(P) = 0.0916 \times P + I(0) \tag{6.3}$$

with $I(0) = 227$ nA. Notice that in the examined range of radiation power, no saturation effect is reached.

Finally, one should remember that Fig. 6.9 emphasizes the high selectivity of pR samples to green light (532 nm). As a matter of fact, the photocurrent increases about 30% with respect to blue light and about 80% with respect to red light. One should also notice that 532 nm is the wavelength to which the maximum of absorption of the pR native state [Friedrich et al. (2002)] corresponds, to be compared with the wavelength of 570 nm corresponding to the maximum absorption for the bR native state [Motto (1980)]. The blue response of pR samples could be associated with the coexistence of proteins in the activated M-state (maximum absorption at 410 nm) and of proteins in the native state, as suggested by recent models of protein dynamics [Alfinito et al. (2011c); Kobilka and Deupi (2007)].

Overall, relevant results of pR experiments can be summarized as follows.

- Due to the presence of pR, there is a significant increase in the current response (for about 27 nA at 120 V and 10 OD). This is an important result since the distance between the electrodes is so large that direct tunneling in a single step becomes impossible and other mechanisms of charge transfer should be considered. Accordingly, charge transfer involving amino acids inside a single protein is suggested as plausible candidates.
- The excess current associated with pR and measured in dark increases for more than 50% in the presence of green light. The observed photocurrent is most likely due to the reaching of the M-intermediate state, that is, to a conformational change, like in bR.
- The observed photocurrent response increases linearly with the light intensity in the analyzed radiation power range (0–150 mW/cm$^2$).

In the frame of the transport mechanisms suggested for bR, and due to the large similarities between these two proteins, it is justified to conclude that (i) like in bR, sequential tunneling between the neighboring amino acids is the main conduction process also in pR and that (ii) the almost linear response in the I–V characteristics of pR is due to the low value of the applied electric field. As a matter of fact, experiments carried out on a single layer of bR (size of about 5 nm) used average electric field values in the range of 0 to 2 $\times$ 10$^6$ V/cm, while experiments on thin films of pR (length about 50 μm) were performed in the range of 0 to 2.4 $\times$ 10$^4$ V/cm.

Current–voltage measurements were also performed across a complete cell containing pR or bR. These evidenced a quasi-linear current response to the membrane potential (few millivolts), further showing that the photocurrent induced in pR increases with voltage slower than in bR under the same conditions [Friedrich et al. (2002)]. Due to the similar experimental setup, this result should imply a different conformational change in the two proteins.

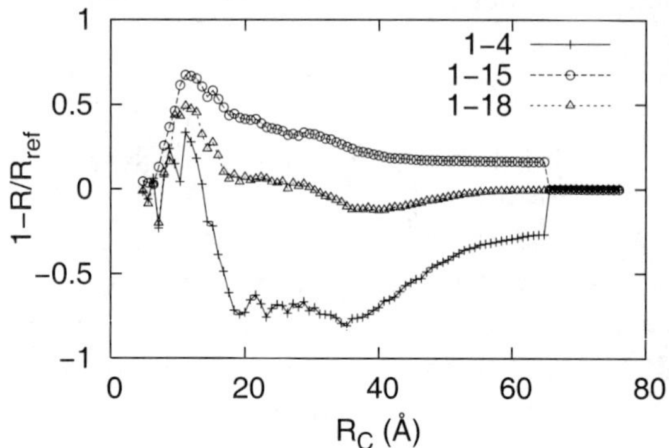


**Figure 6.11** RRV for the native state of the reported pR models, at increasing values of $R_c$. The reference structure is the model number 1 of the PDB entry.

### 6.1.4 *A Comparative Analysis of Proteorhodopsin and Bacteriorhodopsin Electrical Properties*

By construction, the INPA model produces different electrical responses for each of the 20 NMR pR models. Therefore, it is useful to analyze these differences to state the boundaries of theoretical expectations on the basis of the available information.

### 6.1.5 *Protein Resistance*

As a first step, the relative resistance variation (RRV) of the single protein is analyzed for the 20 models of pR, the 2L6X entry of the PDB [Berman et al. (2000)]. Results are reported in Fig. 6.11. Here it is found that there is a peak of differences in the range of interaction radius 6–20 Å. In particular, large deviations correspond to the values of $R_c \sim 10$ Å. One should notice that, although at low $R_c$ values, the resistance of model 1 is larger than that of most of the other models, at greater vales of $R_c$, resistances follow three different pathways:

(i) They can be close to that of model 1 (e.g., model 18 in Fig. 6.11).
(ii) They can be much greater than that of model 1 (e.g., model 4).
(iii) They can be smaller than that of model 1 all over the $R_c$ range (e.g., model 15).

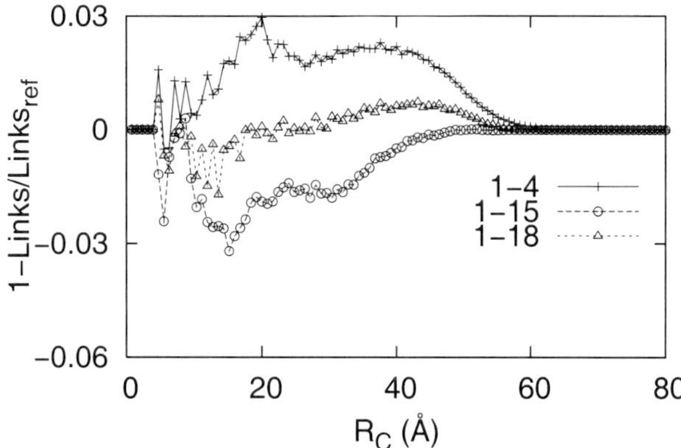


**Figure 6.12** Relative LND for the native state of the pR models, at increasing values of $R_c$. Lines are guides to the eye, and the reference structure is model number 1 of the PDB entry.

The above peculiarities can be verified by analyzing the relative variation of the link number difference (LND), as reported in Fig. 6.12. This outcome suggests the following description of the protein structures:

(i) Models that preserve the relative resistance behavior (with respect to model 1) exhibit a helix distribution similar, and more compact, to that of model 1.
(ii) Models that change the relative resistance behavior exhibit a deformed helix distribution, the larger the higher $R_c$ values.

Finally, Fig. 6.13 shows the RRV between the bR model 2NTU [Berman et al. (2000)] and some pR models (1, 4, 15, 18). To take into account the different protein sizes, the comparison is performed by reporting the normalized resistance values. Calculations exhibit a resistance of bR smaller than that of pR for almost all the models, and more significant for $R_c$ values lower than about 15 Å, i.e., in the most sensitive range.

### 6.1.5.1 Small-signal electrical properties

As reported in Chapter 3, the Nyquist plots well represent the small-signal electrical response of a given protein [Alfinito et al. (2008,

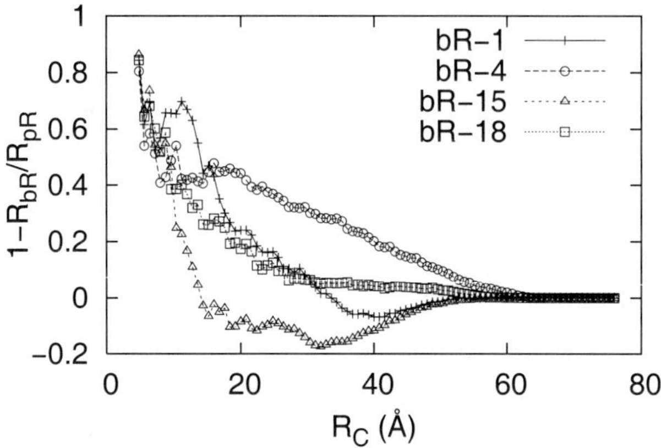


**Figure 6.13** RRV for the native state of the reported pR models, at increasing values of $R_c$. The reference structure is the native state of bR, the entry 2NTU of the PDB [Berman et al. (2000)].

2013a,b)]. Accordingly, Fig. 6.14 reports the theoretical Nyquist plots calculated for the native state of pR and bR single proteins at different $R_c$ values. The shape of the Nyquist plot is found to well resemble the shape of a single RC equivalent circuit also for small $R_c$ values [Alfinito et al. (2008, 2013a)]. One can also observe that the differences among the pR models are quite small for low values of $R_c$ and become significantly larger for high values of $R_c$. Furthermore, at low $R_c$ values, the Nyquist plots of bR are very different with respect to those of pR, while at increasing $R_c$ values, this difference becomes less significant and rather comparable with differences among the considered pR models. It is concluded that the comparison between pR and bR models exhibits the strongest contrast for low $R_c$ values.

### 6.1.5.2 Current–voltage characteristics

In reference [Alfinito et al. (2011a)], it was assumed that the *bona fide* native conductance of a bR single protein was the value deduced from conductive atomic force microscopy (c-AFM) experiments [Casuso et al. (2007a)]. Therefore, the parameters entering the INPA model were calibrated in such a way as to reproduce these

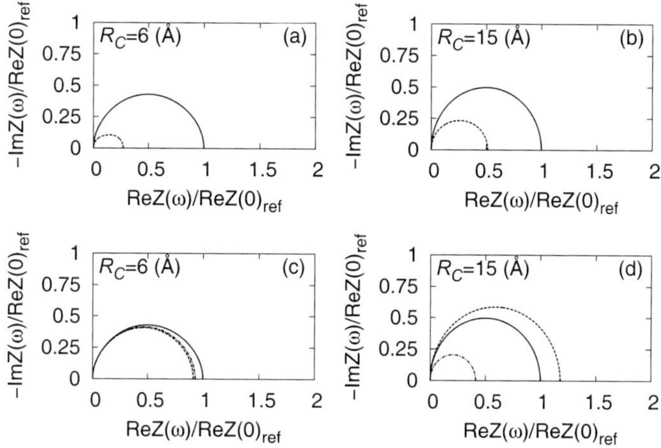


**Figure 6.14** Nyquist plots for the native states of pR model number 1 (continuous line) and bR (dashed line), with $R_c = 6$ Å [panel (a)] and with $R_c = 15$ Å [panel (b)], respectively. Nyquist plots for the native state of pR model number 1 (continuous line), model number 15 (dot-dashed line), model number 4 (dotted line), with $R_c = 6$ Å [panel (c)] and $R_c = 15$ Å [panel (d)], respectively.

**Table 6.1** Theoretical values of bR and pR single-protein conductance, $g$, and conductivity, $\sigma$, as predicted by the INPA model

| Protein | State | $g$(pS) | $\sigma$(S/cm) |
|---|---|---|---|
| bR | Dark | 0.24 | $4.80 \times 10^{-7}$ |
| bR | Light | 0.27 | $5.40 \times 10^{-7}$ |
| pR | Dark | 0.08 | $1.60 \times 10^{-7}$ |

experimental data. In particular, the maximal/minimal resistivity was taken as $\rho_{max} = 4 \times 10^{13}$ Ω Å, $\rho_{min} = 4 \times 10^{5}$ Ω Å, and the barrier height of sequential-tunneling processes $\Phi$ =219 meV.

Table 6.1 reports the single-protein conductance, $g$, both for pR and bR, as calculated within the INPA model. It also reports the conductivity, $\sigma$, associated with a 5 nm side cube with conductance $g$. To calculate the conductivities of native/activated

**Table 6.2** Proteorhodopsin conductivity as inferred from experiments by [Melikyan et al. (2011)]

| OD | Protein | State | $\sigma$ (S/cm) |
|---|---|---|---|
| 10 | pR | Dark | $1.92 \times 10^{-7}$ |
| 10 | pR | Light | $2.80 \times 10^{-7}$ |
| 1 | pR | Dark | $8.74 \times 10^{-8}$ |
| 1 | pR | Light | $1.47 \times 10^{-7}$ |

states of bR and the native state of pR, the PDB files [Berman et al. (2000)] 2NTU, 2NTW, and 2L6X were used. The conductivity value of pR was estimated as the mean value over the 20 structures provided by the PDB entry 2L6X. Accordingly, due to the different tertiary structure, the Ohmic conductivity of native bR was found to be higher than that of the corresponding pR for about a factor of three. The values above should be compared with the experimental values obtained from the literature, as reported in Table 6.2. Data reported in Table 6.2 are obtained after subtracting the buffer conductance (see Fig. 6.10) and are found to be in good agreement with those estimated by INPA calculations. As a further consideration, one should notice that the photocurrent effect in pR is of about 68% at a concentration of 1 OD and decreases to about 50% at 10 OD. Such a decrease could be explained with a stratified growth of the pR protein on the film surface, which shields the effect of impinging photons.

Figure 6.15 reports the theoretical I–V characteristics of bR and pR single protein in their native state. Data are calculated within the INPA model and include tunneling mechanism for charge transfer [Alfinito et al. (2011c, 2013b)]; in particular, each point of the pR curve represents the mean value carried out over the considered 20 models.

Some experiments were performed on samples containing large numbers of bR proteins [Jin et al. (2007, 2006); Ron et al. (2010)], with sample volumes of the order of $10^{-9}$ cm$^3$, and very low bR concentrations. In all these experiments, the protein conductivity inferred from experiments is definitely smaller than that reported above and is in the range (with or without light) of $10^{-11}$–$10^{-13}$

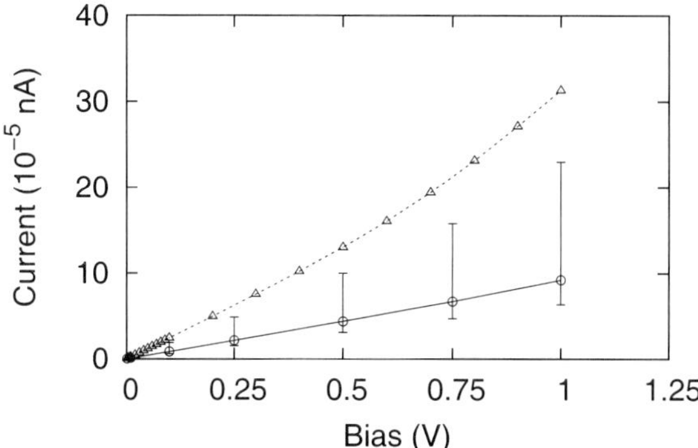


**Figure 6.15** Theoretical I–V characteristics of a single protein in its native state, including sequential-tunneling mechanism of charge transfer. Symbols refer to calculations, and lines are guides for the eyes. The continuous line refers to pR and the dashed line to bR, respectively [Alfinito et al. (2013b)].

S/cm. Nevertheless, the observed photocurrent effect at $10^{-3}$ OD and 1 V is of about 100% [Jin et al. (2006)] in reasonable agreement with the results on pR of reference [Melikyan et al. (2011)].

Figure 6.16 reports the measured photocurrent as a function of the average applied electric field, calculated as the ratio between the applied voltage and active lengths of 5 nm for bR [Jin et al. (2006)] and of $5 \times 10^{-3}$ cm for pR, respectively. The continuous curve refers to the results obtained by the INPA modeling on a single protein of bR, the dashed curve is a linear interpolation of experiments carried out on a thin film of pR. The quasi-linear increase in the photocurrent with applied field is due to the quasi-Ohmic behavior of the I–V characteristics.

To investigate the deviation from the linear Ohmic response at increasing applied fields, Fig. 6.17 reports the chord conductance, I/V, of pR and bR in their native states, as a function of the average applied electric field. Symbols refer to the light-induced responses available from the literature [Jin et al. (2006); Melikyan et al. (2011)] and dashed curves to the results obtained by the INPA single-protein modeling. To take into account the macroscopic size of the samples, data by references [Jin et al. (2006)] and [Melikyan et al. (2011)] are fitted by changing the maximal resistivity from $4.0 \times 10^{13}$ to

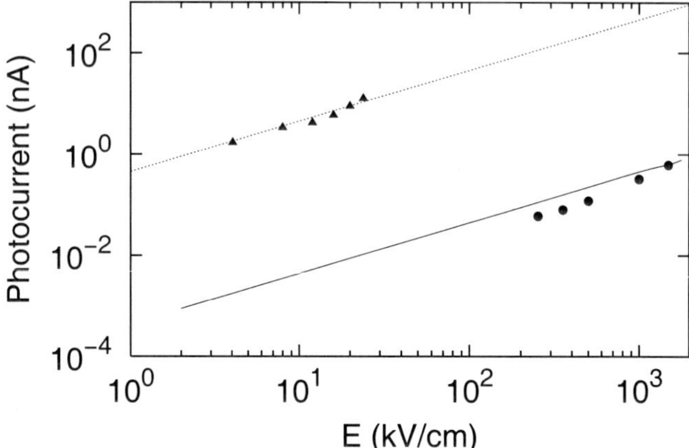


**Figure 6.16** Photocurrent as a function of the average applied electric field. Full circles and triangles represent experimental results for bR [Jin et al. (2006)] and pR [Melikyan et al. (2011)], respectively. The continuous line refers to results of simulations carried out within the INPA model, and the dashed line refers to a linear fitting of pR experiments.

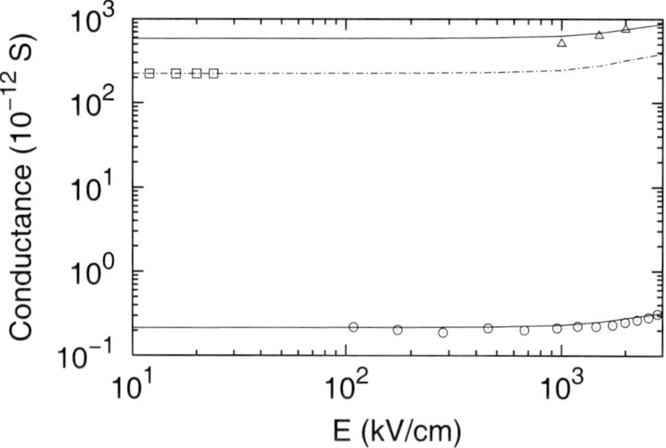


**Figure 6.17** Macroscopic protein conductance under dark conditions as a function of the applied electric field. Symbols refer to macroscopic experimental results, circles for reference [Casuso et al. (2007b)], squares for reference [Melikyan et al. (2011)], triangles for reference [Jin et al. (2006)], lines to simulations carried out by INPA modeling for a single protein. Calculated results are fitted to data obtained at 1 V [Casuso et al. (2007b); Jin et al. (2006)], and to data at the lowest fields [Melikyan et al. (2011)].

$1.6 \times 10^{10}$ Ω Å in the former case and from $4.0 \times 10^{13}$ to $1.5 \times 10^{11}$ Ω Å in the latter case. The comparison between experiments and theory was forced to agree at low fields for pR and at the maximum field for bR. For both the proteins, a deviation from the Ohmic behavior, associated with a transport driven by a field-dependent tunneling mechanism, is predicted at electric fields above about 1 MV/cm. This prediction is in good agreement with experiments carried out on bR nanosamples [Alfinito et al. (2011a); Casuso et al. (2007a,b); Jin et al. (2006)].

### 6.1.6 *Conclusion*

This section investigated the electrical properties of two light receptors, both pertaining to the family of type 1 opsins, pR and bR, whose sensing action is based on a proton pumping mechanism. In particular, the photosensitive properties of pR were detected using MIM thin film structures. The electrical changes due to different pR concentrations, different intensity, and/or wavelengths of the irradiating light were observed by measuring the real-time photocurrent characteristic of the structure. The results of these measurements were then compared with analogous outcomes on bR. To this purpose, the INPA model is used to carry out a comparative microscopic study of the 3D structures and of the conductive properties of both these proteins as a function of an applied voltage. It was assumed that the charge transfer between neighboring amino acids is the microscopic mechanism that determines the macroscopic electrical properties of samples embedding these proteins. In particular, sequential-tunneling mechanism was suggested as the most responsible of the measured current–voltage characteristics. As a matter of fact, by shedding on the protein sample some visible light, a photocurrent arises with a maximum centered at the green region of the spectrum. By analogy with the case of bR, also in pR the presence of a net photocurrent is associated with the conformational change in the single protein due to the presence of light. The results of calculations confirmed these expectations for the electrical properties of the native state of pR. As experimentally found for bR and successfully described within the INPA model, the single-protein conductance of pR was determined and the small-signal response analyzed in terms of the Nyquist plot. Furthermore, also pR is predicted to exhibit super-linear current–voltage characteristics for applied electric fields above about 1 MV/cm. However, the lack of knowledge of the 3D structure of the activated state of pR does not allow to carry out a quantitative microscopic study of its photoconductance.

Those findings are of importance to better understand the basic properties of transmembrane proteins, whose functioning is essential for the living of a single cell. From an applied point of view, opsin-based biomolecular electronics is a valuable premise for the development of a new generation of bio-devices. In particular, it was shown that pR, like bR, represents a relatively simple and stable biological system for exploring charge transfer, since both proteins pertain to a biological material that can be manipulated with all the tools known to modern biophysics. The reported experiments clearly show the sensitivity and usefulness of the MIM technique when used on photoreceptors and open a new scenario for future applications in the wide field of biosensors.

## 6.2 Bovine Rhodopsin

This section investigates the structural and electrical properties of bovine rhodopsin (BR) in dark (native state) and light (activated state). Then a comparison is carried out on the electrical responses before and after undergoing a conformational change by using the INPA approach within the two centers approach, also called AA and AB models, as already anticipated in Chapter 4.

Bovine rhodopsin is one among the most explored structures in the class of G protein–coupled receptors (GPCRs) and is widely used as a prototype for deducing the structure and function of all the other proteins pertaining to the same class of GPCR [Gether and Kobilka (1998); Menon et al. (2001)]. Starting from the atomic coordinates, as reported in the PDB [Berman et al. (2000)], or obtained with different procedures (see Section 6.2.2), first the topological properties of the networks are investigated as a function of the interaction radius between amino acids. Then an analogous investigation is carried out for the global impedance spectrum of the corresponding impedance network.

### 6.2.1 *Modeling*

As anticipated in Section 2.2 of Chapter 2, BR is a seven-$\alpha$-helix transmembrane protein, acting as a light receptor in mammals. It absorbs photons, producing a detecting cascade process that starts with the activation of a G protein and ends with the transmission of information to the brain of the mammal. To date in the PDB [Berman et al. (2000)], there are 26 entries related to BR and less than 20% of these refer to the protein in light. The most used experimental method to obtain the 3D structure is through X-ray, while about 30% of the total entries were obtained with the NMR technique.

The first task addressed here is to assess the level of resolution the present topological network model can reach when discriminating among similar (native or activated) representations and between the native and activated states of a given structure. In other words, one should check whether the network provides a sensitive map of the protein structure. To this purpose, analysis is performed using the AA model.

Figure 6.18 reports the LND in the number of links between couples of BR structures, native (left) and activated (right), as a function of the interaction radius. The comparison is performed among structures obtained from the PDB (native states: 1U19, 2G87, 2HPY, and activated states: 1LN6, 2I37), and from homology models (see Section 6.2.2). In particular, native (Rho) and activated (Meta II) configurations refer to engineered structures obtained from the incomplete structures of activated state, 1LN6, and native states, 1JFP and 1F88 (chain A). The three native representations of BR (1U19, 2G87, and 2HPY) were compared with the engineered one, Rho. Then the main results are as follows: (i) 1U19 and 2G87 exhibit practically the same number of links and (ii) there are consistent differences between the native and engineered configurations. Furthermore, when moving from the native to the activated state, here represented by structures obtained with three different methods—Meta II (homology modeling), 1LN6 (NMR), and

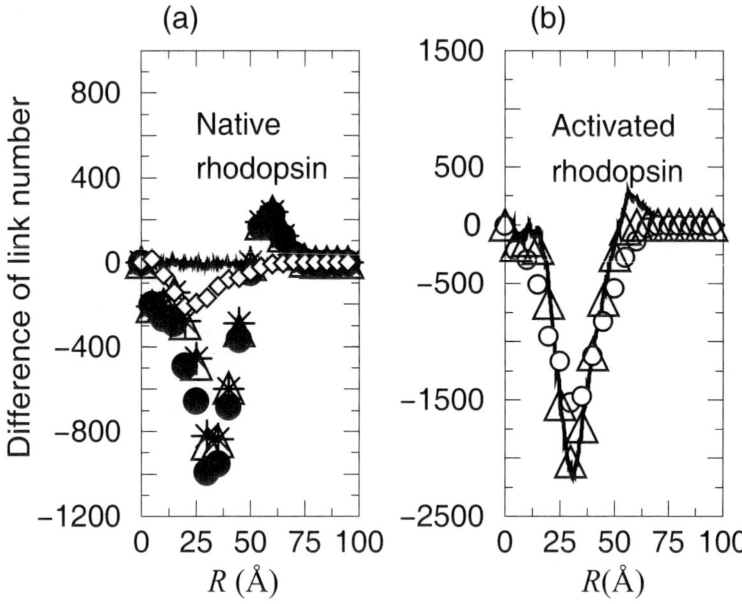


**Figure 6.18** LND for different configurations of BR versus the interaction radius. All the configurations contain the same number of amino acids. Figure (a), Native BR: Stars refer to the LND between the chain A of 1U19, 1U19*a*, and the engineered representation of native BR, Rho. Open triangles refer to the LND between 2G87 and Rho. Full circles refer to the LND between 2HPY and Rho. Open diamonds refer to the LND between 2HPY*a* and 1U19*a*. Continuous line refers to the LND between 2G87 and 1U19. Figure (b), Activated BR: Continuous line refers to the LND between the engineered representations of BR in light, Meta II, and Rho. Open circles refer to the LND between 2I37 and Rho. Open triangles refer to the difference between 1LN6 and Rho. In the second and third cases, the Rho sequence has been deprived of the amino acids that are not present in 2I37 and 1LN6, respectively [Alfinito et al. (2008)].

2I37 (X-ray)—an increase in the difference in the number of links is found.

From Fig. 6.18, it is evident that the representations in the same state (native or activated) look sometimes comparable with those of the representations of the same protein in different states. In other words, different experimental conditions may produce very different representations of the same state of the protein. Therefore,

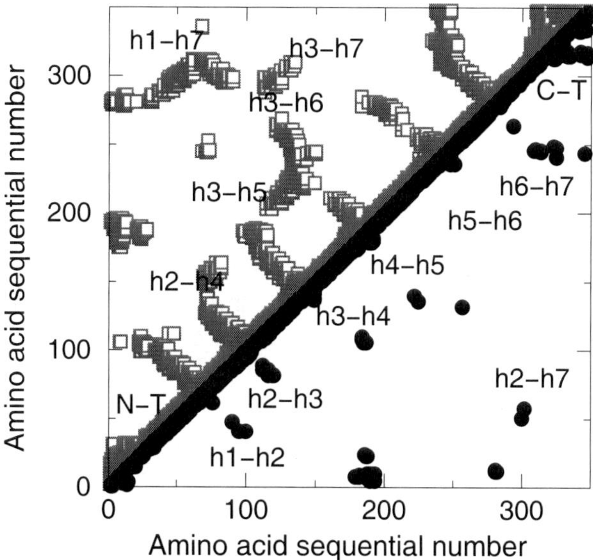


**Figure 6.19** Contact map of native BR, Rho. The $x$- and $y$-axes report the sequential number of the amino acids. Full circles refer to $R_c = 6$ Å, and open squares to $R_c = 12$ Å. Each circle/square corresponds to a link between the couple of amino acids $(x,y)$. The main domains associated with the connections among closest helices are explicitly indicated.

if the network model is used to discriminate between the native and the activated (or complexed) states of the protein, it is mandatory that the reference representations be produced under the same experimental conditions. For this reason, in the following the couple Rho-Meta II is used for native and activated BR.

To emphasize the ability of the network model to capture the protein topology, Fig. 6.19 reports the contact map of BR in dark, Rho. The helix-to-helix (h♯-h♮) links are very evident, either for $R_c = 6$ Å (dark points) or, even better, for $R_c = 12$ Å (grey boxes). It is found that the links reproduce the closeness of h2 with h3 and h1, and of h2 and h4. Furthermore, they suggest the presence of H-bonds among h1, h2, h7 and h2, h3, h4 and also between h3 and h6 and between h6 and h7 [Menon et al. (2001)]. Notice that the helix couples (h1-h3), (h1-h4), (h1-h5), (h1-h6), (h2-h5), (h2-h6), (h2-h6), (h4-h7), (h5-h7) are not connected for these values of the interaction radius. It is concluded that the increase in $R_c$ is equivalent to interacting forces with longer range. Accordingly, the drawings in Fig. 6.19 emphasize the dependence from the interacting radius of the connective map of the network.

### 6.2.2 *Engineering of Bovine Rhodopsin Spatial Structure*

In the absence of a complete sequence of amino acids, the full 3D structures of BR were obtained from a combination of two data sets as described below. BR structures reported in PDB are assumed to be relatively isomorphic. Thus, the superposition of their coordinate systems, obtained by successive translations and rotations, is sufficient for inserting one data set into another one. The way of the superposition is as follows. The coordinates of three $C_\alpha$ positions present in both data sets, say, A1, B1, C1 and the corresponding A2, B2, C2, are superposed. Specifically, the coordinates of A1 and A2 are matched by translation, then the directions of vectors (A1-B1) and (A2-B2) are made coincident by using the first rotation; finally the planes (A1-B1-C1) and (A2-B2-C2) are superposed with the second rotation. No scaling is used since chemical bonds should have fixed length. To estimate the correctness of a compiled structure, the average distance between corresponding $C_\alpha$ atoms of processed data sets, $< \Delta >$, is calculated. The full data set for the native state of BR is obtained from the data set 1JFP by adding their 39 amino acids from the data set 1F88 (chain A). The amino acids missed in the 1JFP set form a continuous part of backbone chain with amino acid numbers from 1 to 39. Thus, in the protein backbone, there is only one place where it is possible to expect local misfits, caused by the nonisomorphic shape of 1JFP and 1F88(A). Accordingly, to avoid this misfit, amino acid number 40 was chosen as point A. Then the $< \Delta >$ value was numerically minimized over all possible combinations of B and C. The resulting $< \Delta >$ value is $\Delta_{min} = 5.5$ Å, which is a reasonable value by considering that the characteristic BR size is about 50 Å. The native state corresponding to $< \Delta >_{min}$ is formed by amino acids with numbers 40, 155, and 282, corresponding to A, B, and C. To construct Meta II data set from 1LN6 and 1F88(A), the same procedure was followed, with the same basis. 1LN6 was engineered on the base of 1JFP [Choi (2002)] and, therefore, the same amino acids were missed there. The missed amino acids all belong to the N-terminus of the protein, while the fundamental differences between Rho and Meta II states are supposed to be in the transmembrane core.

To date, several 3D structures pertaining to BR sets are available in the PDB, some of which are complete. Furthermore, some laboratories offer the possibility to produce 3D protein models by using the primary structure.

### 6.2.3 *Small-Signal Electrical Properties*

Here the frequency response of the impedance network, $Z(f)$, in the range 0–1100 Hz, is calculated and represented by the Nyquist plot for the BR single protein in its native and activated states and for different values of the interaction radius.

The INPA AA model contains one free parameter in the value of the cut-off radius $R_c$, which fixes the number of links and so the network topology. In the limit of small $R_c$ values (to say $R_c <$ 6 Å), only the nearest neighbors are connected, and so it is not possible to reveal the existence of more complex structure like $\alpha$-helices or $\beta$-sheets. By contrast, in the limit of large $R_c$ values (to say $R_c >$ 80 Å), each node is connected with all the others, and so the protein appears as a uniform structure. A relevant value of $R_c$ for the purposes of emphasizing the effects associated with the conformational change should enable the main structures of the protein to emerge clearly. Indeed, one is interested in detecting if and how they displace conformational change in the protein. Accordingly, one should look for a value of $R_c$ best revealing the main structures of the protein but also emphasizing the differences between the activated and the native states of the protein. For a GPCR, a relevant value of $R_c$ is a compromise between the characteristic dimension of the $\alpha$-helices and the typical distance among $\alpha$-helices, say $D$ [Alfinito et al. (2008)]. When BR goes from the native to the activated state, its $\alpha$-helices change their relative distance and $D$ goes in $D'$. In the corresponding network, when the value of $R_c$ is between $D$ and $D'$, a relevant number of links change their value, so well revealing the conformational change. $D$ is thus called the *effective distance* and is used as a reference length value.

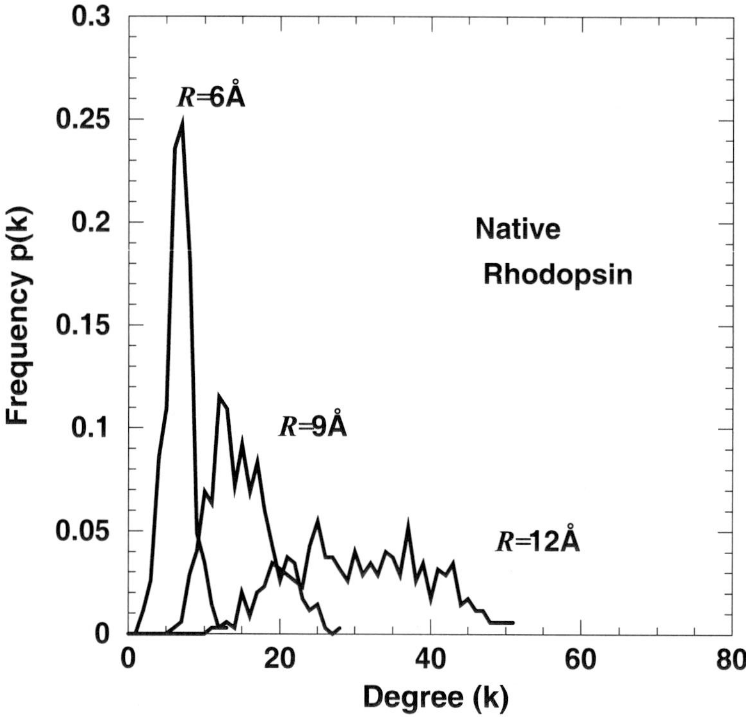


**Figure 6.20** Degree distribution of native BR (Rho) network for increasing values of the interaction radius $R_c$.

To explore the different topologies associated with the changing of $R_c$, in the frame of the AA model, it is appropriate to evaluate the network degree distribution, i.e., the distribution of the connected nodes [Albert and Barabási (2002)]. To this purpose, Fig. 6.20 reports the results of calculations. Here, for $R_c \leq 9$ Å, one observes that the degree distribution remains substantially peaked around the same degree value, i.e., there is a single characteristic dimension of the network clustering. It corresponds to the nearest neighboring domain $k = 7$. The cluster dimension grows for increasing $R_c$, up to the value $R_c = 12$ Å. Indeed, one notices that for $R_c = 12$ Å the degree distribution of BR exhibits two prominent maxima at $k = 25$ and 37, respectively (see Fig. 6.20), and so two different clusterizations are found. For values of $R_c$ in the range of 12–25 Å, the distribution exhibits a spreading, which shows a series of spikes representing the fingerprint of the tertiary structure. For values of $R_c > 25$ Å, the degree distribution is found to shrink and at $R_c = 80$ Å, all the nodes are found to be practically connected to each other. Here the degree distribution takes a delta-like shape centered at $k = (\nu - 1)$. It is concluded that $R_c = 12$ Å for BR should be taken as optimal values to obtain the best resolution of the intimate protein structures.

From the above considerations, in the following are discussed four possible cases, in which the interaction radius $R_c$ and the effective distance $D$ combine to produce different resolutions for the AA- and AB-directed models, respectively. (Notice that the effective distance for the AB model is taken to be larger than that for the AA model because of the finite size of the amino acids.)

$$\begin{array}{lllll} I. & D_{AA} \approx R, & D_{AB} \gtrsim R, & D'_{AA} \gtrsim R, & D'_{AB} \gtrsim R \\ II. & D_{AA} \lesssim R, & D_{AB} \approx R, & D'_{AA} \gtrsim R, & D'_{AB} \gtrsim R \\ III. & D_{AA} < R, & D_{AB} \lesssim R, & D'_{AA} \lesssim R, & D'_{AB} \gtrsim R \\ IV. & D_{AA} \ll R, & D_{AB} < R, & D'_{AA} \ll R, & D'_{AB} < R \end{array} \quad (6.4)$$

Since, contrary to the condition $D > R$, the condition $D < R$ produces more links, the previous cases are analyzed as follows:

Case I. Here the AA model discriminates different protein states better than the AB model.

Case II. Here both the AA and the AB models are able to resolve well the two configurations. In particular, the AA model is more sensitive to the change of the interaction radius.

Case III. Here the AA model discriminates different protein states worse than the AB model.

Case IV. Here it is rather difficult to discriminate the configurations both for the AA and the AB models since the number of links remains practically the same in both the configurations.

The general trends discussed above can be quantitatively assessed for the proteins under test by selecting for BR the significative set of $R_c$ values 6, 12, 25, and 50 Å. The numerical results of calculations are reported in Appendix with an accuracy of three digits, which is considered appropriate for an experimental validation of the model.

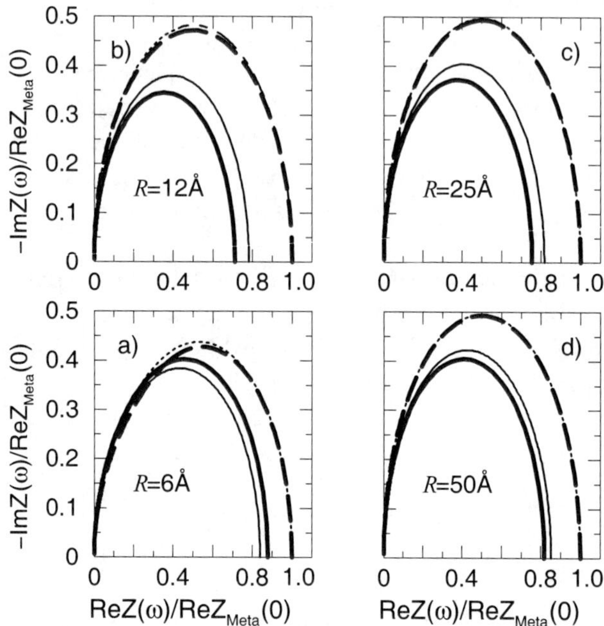


**Figure 6.21** Nyquist plots of the network impedance associated with the protein. Native BR is compared with activated BR. The protein sequences are engineered Rho (native) and Meta II (activated). For the AA model, continuous tiny lines refer to Rho, and dotted lines refer to Meta II. For the $AB_{\alpha,\alpha}$-directed model, bold continuous lines refer to Rho, and dashed lines refer to Meta II. Plots are reported for increasing values of the interaction radius in the range 6–50 Å following the clockwise orientation.

Figures 6.21–6.23 report the Nyquist plots of the global network impedance, normalized to the value at zero frequency, $Z(0)$, for the case of the engineered representations of BR in the native state, Rho, and in the activated state, Meta II. In all the figures, the AA model is compared with the AB model by adopting the same convention for the used symbols. In each figure, the plots corresponding to increasing values of $R_c$ are indicated as (a), (b), (c), and (d) in the clockwise orientation. As a general trend, the shape of the Nyquist spectra remains quite close to that of a semicircle, typical of a single RC parallel impedance, except for small but significant deviations from the semicircle when $R_c = 6$ Å.

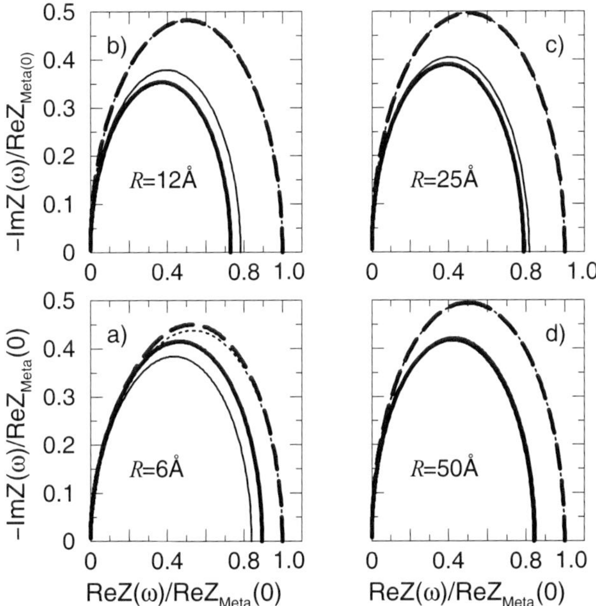


**Figure 6.22** Nyquist plot of the network impedance associated with the protein. Native BR is compared with activated BR. The protein sequences are engineered Rho (native) and Meta II (activated). For the AA model, continuous tiny lines refer to Rho, and dotted lines refer to Meta II. For the $AB_{\alpha\beta,\alpha\beta}$-directed model, bold continuous lines refer to Rho, and dashed lines refer to Meta II. Plots are reported for increasing values of the interaction radius in the range of 6–50 Å, following a clockwise orientation.

Figure 6.21 shows the different impedance responses obtained with the AA model (tiny continuous line for Rho and dotted line for Meta II) and with the $AB_{\alpha,\alpha}$-directed model (bold continuous line for Rho and dashed line for Meta II).

Figure 6.22 shows the Nyquist plots for the AA model and the $AB_{\alpha\beta,\alpha\beta}$-directed model, respectively.

Figure 6.23 shows the Nyquist plot for the AA model and the AB isotropic model.

For $R_c = 6$ Å, in all the cases the AA model exhibits a resolution between different configurations better than the AB models. Thereby the value $R_c = 6$ Å pertains to case I: This value is a good choice but not the best. The value of $R_c = 12$ Å provides the

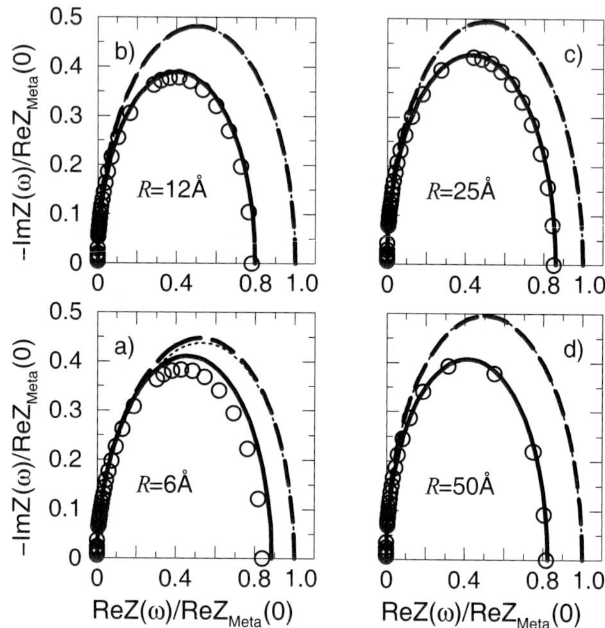


**Figure 6.23** Nyquist plot of the network impedance associated with the protein. Native BR is compared with activated BR. The protein sequences are engineered Rho (native) and Meta II (activated). For the AA model, circles refer to Rho, and dotted lines refer to Meta II. For the AB isotropic model (with in-contact on the first $C_\alpha$ carbon atom and out-contact on the last $C_\alpha$ carbon atom), bold continuous lines refer to Rho, and dashed lines refer to Meta II. Plots are reported for increasing values of the interaction radius in the range of 6–50 Å following a clockwise orientation.

largest difference between the activated configuration and the native one. Furthermore, it is found that the $AB_{\alpha,\alpha}$-directed model (see Fig. 6.21), and the $AB_{\alpha\beta,\alpha\beta}$-directed model (see Fig. 6.22) increase the differences, respectively, of 7% and 5% with respect to the AA model. The value $R_c = 12$ Å pertains to case II; the best value for the AA model, a very good value for the AB-directed model. For $R_c = 25$ Å, the difference between the configurations starts to decrease for both the models, even if the directed AB model still exhibits a resolution increment of 7% with respect to that of the AA model, and 3% for its versions $AB_{\alpha,\alpha}$ and $AB_{\alpha\beta,\alpha\beta}$, respectively. The value $R_c = 25$ Å pertains to case III: For the AB models, this value of $R_c$ is still a relevant one. Finally, for $R_c = 50$ Å, the directed AB model exhibits a resolution increment of 3% with respect to that of the AA model, and 1% for its versions $AB_{\alpha,\alpha}$ and $AB_{\alpha\beta,\alpha\beta}$, respectively. The ability to resolve is decreasing, but it remains significant for both the models. Accordingly, the value of $R_c = 50$ Å is on the boundary between case III and case IV.

Figure 6.23 reports the comparison between the AA and the isotropic AB model. One can observe that for $R_c > 6$ Å the compared models give practically the same results; in other words, the AB isotropic model does not improve the AA model. This outcome means that the AB isotropic model does not give to the $C_\beta$s an active role, unlike the AB-directed model. With respect to the directed model, it contains much more links, many of them slightly varying in the conformational change. This excess of (invariant) links hides the small differences between the native and activated states. Therefore, while the larger effects due to $C_\alpha$ displacements can emerge once again, the smaller improvements due to the $C_\beta$ displacements cannot be appreciated.

In Fig. 6.22 and Fig. 6.23, one can appreciate that for $R_c = 6$ Å, the Nyquist plots take shapes that are slightly squeezed and asymmetric semicircles instead of the perfect semicircle pertaining to a single RC parallel circuit (for details, see Appendix). However, by increasing the value of $R_c$, the Nyquist plots better and better approach the perfect semicircle shape. The above peculiarities can be satisfactorily interpreted in terms of the Cole–Cole function [Cole and Cole (1941)] with one fitting parameter. Accordingly, to interpret the Nyquist plot, the normalized dimensionless response function [Macdonald (1987)] is used:

$$I_\omega \equiv (Z(\omega) - Z_\infty)/(Z(0) - Z_\infty) \tag{6.5}$$

which reduces to $Z(\omega)/Z(0)$ in this case.

The Cole–Cole fitting function is:

$$I_\omega = \frac{1}{1 + (i\omega\tau)^{1-\alpha}}, \qquad 0 \le \alpha \le 1 \tag{6.6}$$

which leads to the following relation between the real and imaginary parts of $I_\omega$.

$$\left(\mathcal{R}(I_\omega) - \frac{a}{2}\right)^2 + \left(\mathcal{I}(I_\omega) - \frac{b}{2}\right)^2 = \frac{1}{c}(1 - a) + \frac{1}{4} \tag{6.7}$$

with

$$a = cos(\pi\alpha/2)$$

$$b = sin(\pi\alpha/2)$$

$$c = 1 + (\omega\tau)^{2(1-\alpha)} + 2(\omega\tau)^{(1-\alpha)}b$$

where $1/\tau = \omega_M$ is the frequency value corresponding to the maximum value taken by $-\mathcal{I}(I_\omega)$ as a function of $\mathcal{R}(I_\omega)$ (see Appendix).

The Cole–Cole function is one of the most used fitting functions in relaxation processes deviating from the Debye–Maxwell behavior [Debye (1929)]. Of course, it is not unique due to the complexity of the possible origins of this deviation [Davidson and Cole (1951); Macdonald (1987)]. However, it has been shown that its meaning is more extensive than that of a simple fitting function [Metzler and Klafter (2002)]. Indeed, while the power spectrum associated with the Debye–Maxwell function, $|I_\omega|^2$, is Lorentzian, and the correlation function is exponential, the power spectrum associated with the Cole–Cole function is a more tangled object. It reduces to the Lorentzian distribution for $\alpha = 0$. Furthermore, it goes like $(\omega\tau)^{-2(1-\alpha)}$ for $\omega\tau \gg 1$ and like $(1 + 2(\omega\tau)^{(1-\alpha)}sin(\pi\alpha/2))^{-1}$ for $\omega\tau \ll 1$. The corresponding correlation function is the Mittag–Leffler function [Abramowitz and Stegun (2012)], which interpolates between a stretched exponential pattern ($\omega\tau \gg 1$) and an inverse power law decay ($\omega\tau \ll 1$). The exponent of both the functions is same, (1-$\alpha$).

For completeness, Appendix reports the single resistance and capacitance values corresponding to the calculated Nyquist plots.

### 6.2.4 *Current–Voltage Characteristics*

Calculations of the I–V characteristics are carried out for the native and activated states of BR, whose structures were engineered as reported in reference [Alfinito et al. (2008)]. The obtained results are reported in Fig. 6.24. Remarkably, for an analogous conformational change, the INPA model predicts that BR exhibits a behavior opposite to that of bR, i.e., in going from the native to the activated configuration, the current at a given voltage decreases.

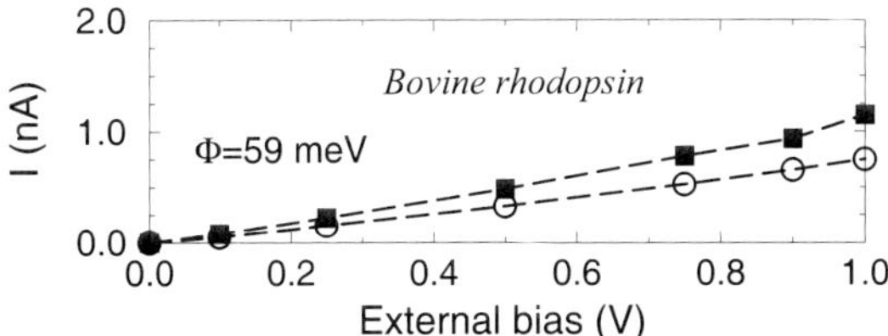


**Figure 6.24** Predicted I–V characteristics of the native (full squares) and activated (open circles) states of BR; dashed curves are guides for the eyes. Calculations are carried out using the same parameters of bR as reported in Section 5.3 of Chapter 5. All the currents obtained by calculations are normalized to the experimental value of current in dark at 1 V [Jin et al. (2006)].

Physically, it is possible that this opposite behavior follows from the opposite change that the retinal undergoes when absorbing a photon. In BR the retinal shape goes from bent to straight, while the reverse occurs in bR [Berman et al. (2000)]. As a consequence, a plausible conclusion is that an opposite conformational change of the entire protein should imply an opposite change of its I–V characteristics.

### 6.2.5 *Conclusion*

This section reports a systematic analysis of the electrical and related properties of BR, the prototype of the GPCR family. By using the topological features of the impedance network, some relevant PDB entries have been compared in terms of the number of links as a function of the interacting radius.

By using the features of the impedance network associated with the topological one, the small-signal impedance of the proteins is reported through the Nyquist plot representation. Accordingly, for BR an electrical response is predicted, which is of a quite detectable level (up to difference of 22% when passing from the native to the activated state). Furthermore, a significative conformational change is identified both with the one-node AA model and with the two-node AB model, the latter foreseeing significantly larger differences among the configurations. These results are supported by some experimental evidences [Hou et al. (2006, 2007)]: BR was immobilized on a gold electrode building up a self-assembled multilayer (SAM). The electrochemical impedance spectroscopy (EIS) characteristic of these structures was carried out in a standard electrochemical cell and has shown Nyquist plots qualitatively similar to those presented in this section.

## 6.3 Rat OR I7

This section investigates the structural and electrical properties of the single-sensing protein rat OR I7 pertaining to the GPCR family. In particular, the change in the impedance spectrum following a conformational change will be analyzed.

### 6.3.1 *Modeling*

As input data of the model, the tertiary structure of the protein in both its native and activated states is needed. First principle calculations, X-ray crystallography and protein NMR suggested the 3D structure of the rat OR I7 native state [Berman et al. (2000); Hall et al. (2004); Vaidehi et al. (2002)]. By contrast, no full information on the activated state is reported in literature. Furthermore, if the acquisition conditions of the two protein morphologies (i.e., native and activated states) are different, then the resulting structures cannot be reliably compared [Alfinito et al. (2008)]. To overcome these difficulties, the tertiary structures of native and activated states for rat OR I7 were obtained by means of the MODELLER software, a program that implements a comparative protein structure modeling by satisfying of spatial restraints [Fiser and Sali (2003); Sali and Blundell (1990)]. The template PDB entries were BR 1JFP and 1LN6 for the native and activated states, respectively, complemented with 1F88, chain A [Berman et al. (2000)].

Figures 6.25–6.26 report the backbone (left panel a) and sphere (right panel b) representations of rat OR I7 in its native and activated (meta) states, as obtained by the homology modeling procedure. These entries were selected because both 1JFP and 1LN6 were produced in the same experimental conditions. On the

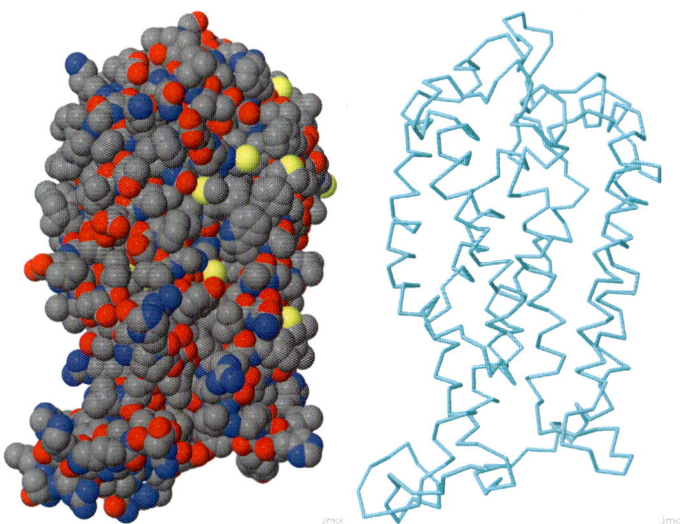

**Figure 6.25** Sphere (left panel a) and backbone (right panel b) representations of native rat OR I7.

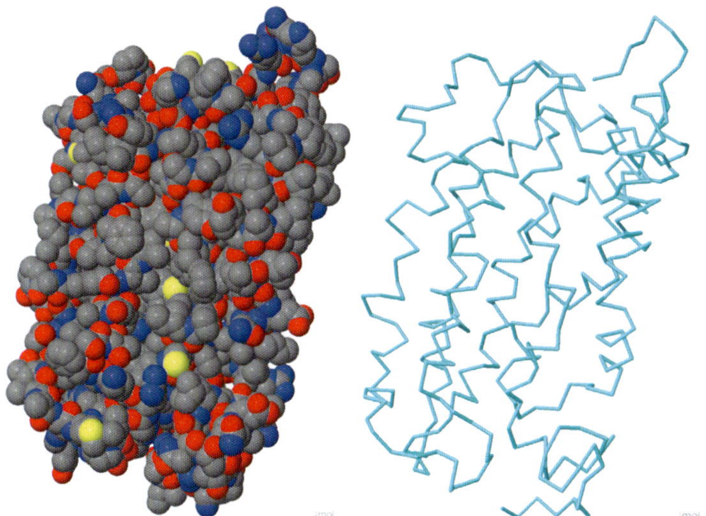

**Figure 6.26** Sphere (left panel a) and backbone (right panel b) representations of active rat OR I7.

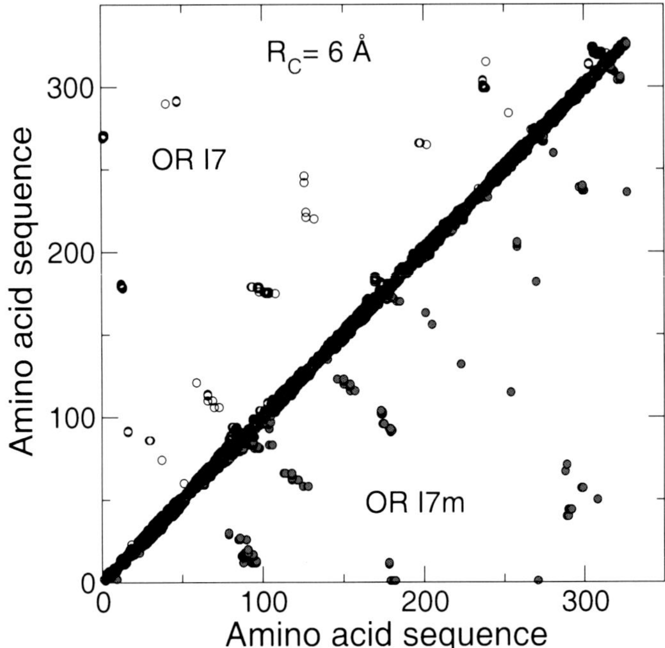


**Figure 6.27** Contact map of the two rat OR I7 representations: native state (I7 and squares on the left) and activated state (I7m and squares on the right). Because of the axial symmetry with respect to the diagonal exhibited by data, the symbols pertaining to the native map are reported only on the left-hand side of the diagonal and those pertaining to the activated map on the right-hand side. The interaction radius is $R_c = 6$ Å.

other hand, quite analogous results were obtained by choosing other entries.

## 6.3.2 *Topological Properties*

Figures 6.27–6.28 report the contact maps for rat OR I7 in the native and activated states calculated with the two different $R_c$ values of 6 and 12 Å, respectively. The figures above show that by increasing the value of $R_c$, the density of points and thus the connectivity increase substantially, as expected. In particular, the main differences between the native and activated states are in the increased connectivity among the helices h3-h4-h6, where

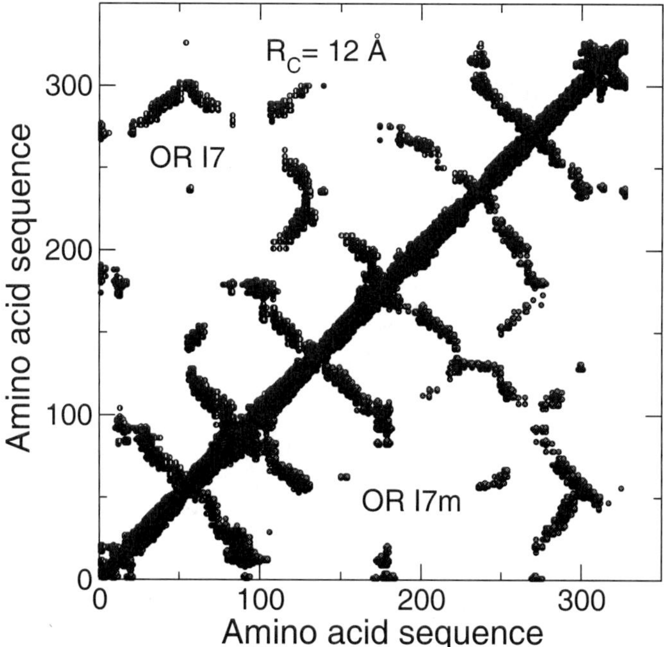


**Figure 6.28** The same as in Fig. 6.27 for an interaction radius of $R_c = 12$ Å.

the principal binding pocket is presumably located [Alfinito et al. (2011c); Hall et al. (2004)]. Notably, significant differences between the native and activated states survive for large $R_c$, say up to $R_c$ values of about 50 Å [Alfinito et al. (2011b)]. The differences between the native and activated states of rat OR I7 are evidenced in the change in the receptor global resistance, as shown in Fig. 6.29, which reports the relative variation of resistance as a function of $R_c$. The main result evidenced by this figure is the large sensitivity of the rat OR I7 with a maximum resistance variation of about 60% for $R_c$ in the range of 8–14 Å.

### 6.3.3 *Small-Signal Electrical Properties*

Figure 6.30 reports EIS spectra obtained for different concentration of rat OR I7 by Hou et al. (2007). The spectra are found to change according to the concentration of the deposited membrane fraction

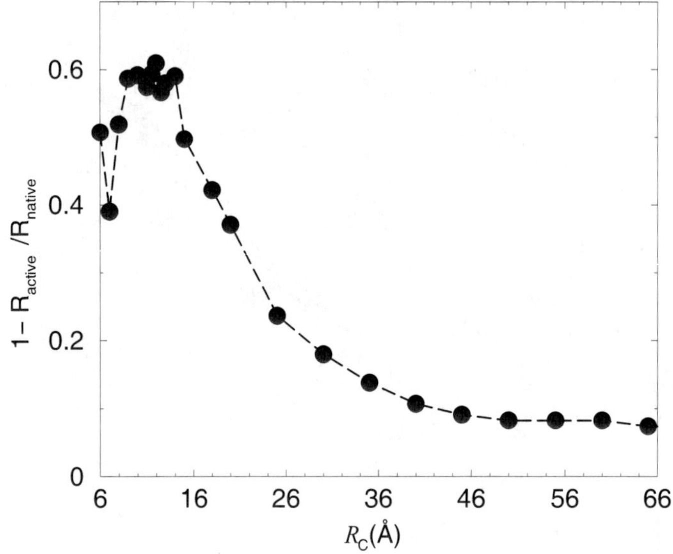


**Figure 6.29** RRV at increasing $R_c$ values for rat OR I7.

on its specific biotin layer antibody, thus confirming that EIS can be a powerful tool to monitor the different stages in the preparation of the sample.

The EIS responses are usefully decoded in terms of a simple equivalent electric circuit, the Randles cell, which fits the EIS experiment using a few number of passive elements, as depicted in the insert of Fig. 6.30. Here $R_S$ is the resistance of the solution filling the electrochemical cell, and $R_P$, $Z_W$, and $Z_{CPE}$ are, respectively, the polarization resistance, the Warburg impedance, and the constant phase element of the Randles cell.

For the rat OR I7, impedance responses able to discriminate specific odorants and their concentration were observed by Hou et al. [Hou et al. (2007)]. The selective odorant detection of this receptor was tested onto two different aldehydes: The OR-specific odorant, octanal (I), and the nonspecific, helional (II). The analysis was performed by measuring the variation of the polarization resistance before, $R_{P,b}$, and after, $R_{P,a}$, the injection of the odorants at different concentrations, as reported in Fig. 6.31.

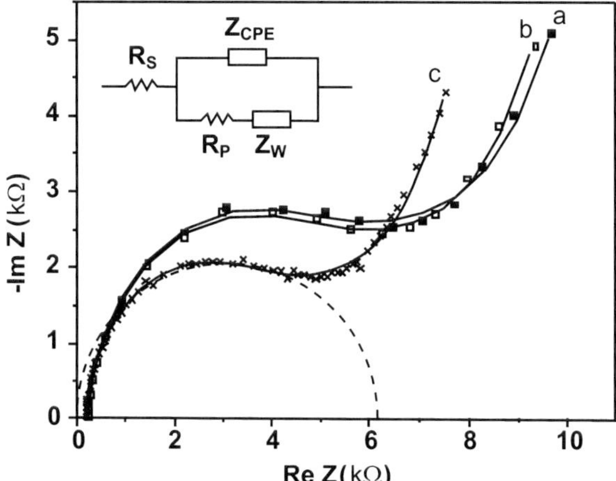


**Figure 6.30** Electrochemical impedance characteristics of rat OR I7 immobilized on its specific biotinylated antibody deposited on a SAM at different protein concentrations: (a) 0 mg/ml, (b) 20 ng/ml, and (c) 150 ng/ml [Hou et al. (2007)]. Symbols refer to experimental data, and continuous lines refer to the fitting performed using the equivalent circuit analogue (Randles cell) depicted in the insert. The dashed semicircle corresponds to the Nyquist plot of an ideal RC parallel circuit.

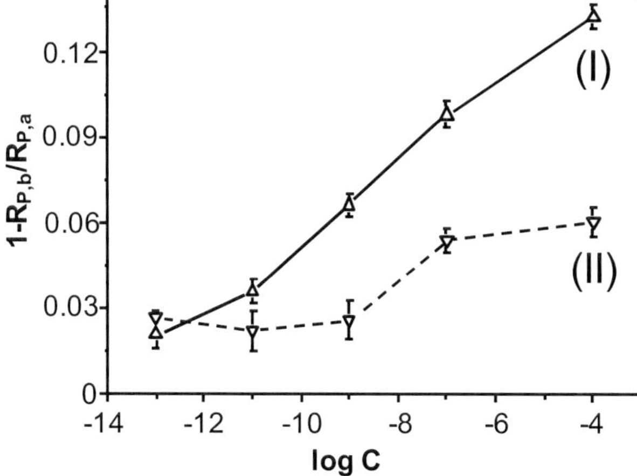


**Figure 6.31** Dose response versus the log of concentration (expressed in molarity) of rat OR I7 to octanal (I), helional (II). $R_{P,b}$ and $R_{P,a}$ indicate the polarization resistance, respectively, before and after the odorant injection.

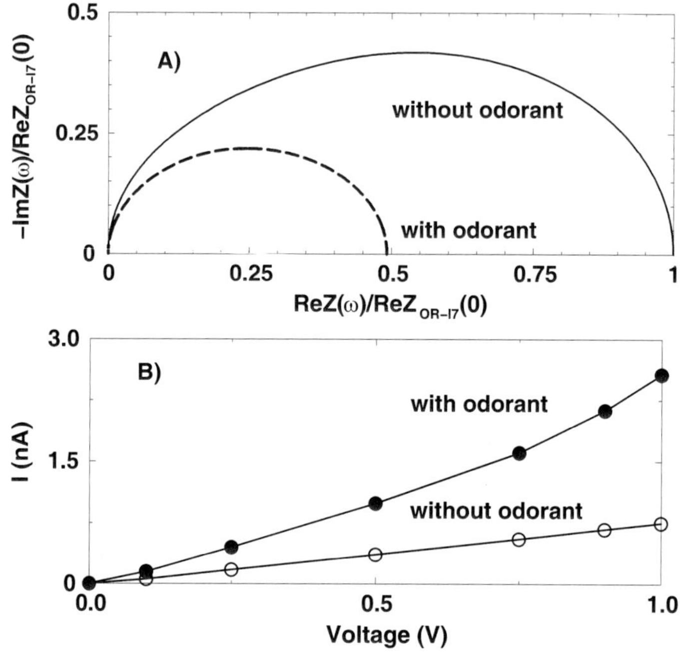


**Figure 6.32** Nyquist plot (panel A) and I–V characteristics (panel B) of the rat OR I7 in the presence or less of the specific odorant.

The net result of these experiments is a decrease in the polarization resistance for increasing odorant concentrations, with a maximum variation of about 15% for octanal, when going from a concentration of $10^{-12}$ M to that of $10^{-4}$ M.

Figure 6.32 part (a) shows the calculated Nyquist plot for the impedance of rat OR I7 in the native and activated states when $R_C = 6$ Å, a value which gives a significant difference between the two states [Alfinito et al. (2009c)]. To focus on the variation of the impedance associated with the conformational change, the real and imaginary parts of the impedance are normalized to the value of the static impedance of the native configuration. The change in impedance associated with the change in conformation for a single protein is found to be within a factor of two, thus easily detectable by experiments. This variation compares favorably with the experimental measurements, as reported in Fig. 6.33.

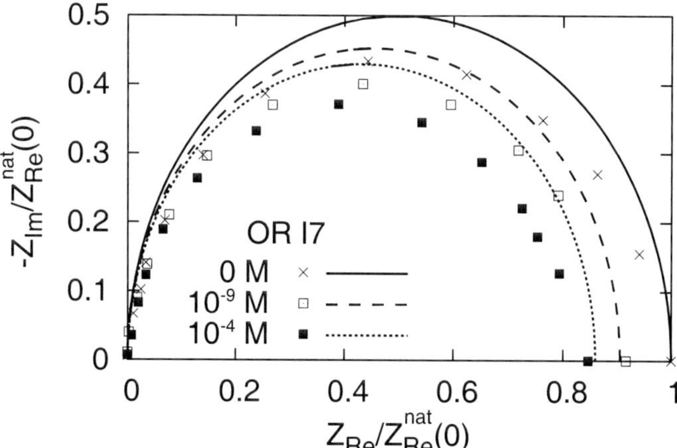


**Figure 6.33** Nyquist plot of rat OR I7 in the presence of different concentrations of the specific odorant. Curves refer to calculations and symbols to experiments.

## 6.3.4 *Current–Voltage Characteristics*

For the sake of completeness, the static I–V characteristic was also calculated corresponding to an OR I7 nanolayer when sandwiched between metallic contacts, in analogy with experiments reported in reference [Jin et al. (2006)] for the case of the light receptor bR. Following Chapter 4, calculations are carried out by assuming the presence of a tunneling mechanism of charge transport at increasing voltages [Alfinito et al. (2009c)] in both the native and activated states. The results of these calculations are reported in Fig. 6.32 panel (b). Here for fixed bias values, a significant enhancement of the current is found when going from the native to the activated state. Therefore, by analogy with the case of bR, this kind of measurement should be a very useful tool to further analyze the sensitivity of the rat OR I7 to specific odorants. The main conclusions that can be drawn from Fig. 6.32 panels (a) and (b) are that rat OR I7 exhibits:

(i) a net change in impedance in going from the native to the activated state;
(ii) an activated state with the static impedance (resistance) being about 50% smaller than that of the native state; and

(iii) a static I–V characteristic with the current pertaining to the activated state significantly higher than that corresponding to the native state.

Remarkably, all the above results pertain to an ideal structure in the absence of thermal fluctuations, and with a saturated concentration of odorant. Nonetheless, by comparing these predictions with the experiments reported in Fig. 6.31 and Fig. 6.33, a good qualitative and quantitative agreement is found for the change in the impedance following the sensing action of rat OR I7. The body of these results clearly indicates that the static and AC electrical responses are powerful tools for monitoring the sensing action of GPCR proteins and OR receptors in particular.

### 6.3.5 *Conclusion*

This section investigates the possibility to detect modifications of the electrical response of a sensing protein belonging to the GPCR family due to the capture of a specific ligand. This ability has roots in a wide set of observable data that support the idea that topology does matter in protein functioning. In particular, evidence of significative modifications was found in the electrical properties of rat OR I7 as a consequence of the protein conformational change. The obtained results support the possibility to develop a nanobiosensor for odorant recognition based on the electrical properties of an olfactory receptor.

## 6.4 Human OR 17–40

This section reports the electrical and related properties of the sensing protein human OR 17–40. In particular, the change in the impedance spectrum following a conformational change is investigated.

### 6.4.1 *Modeling*

Since no crystallographic data are available for the 3D structures of the human OR 17–40, its 3D structures are obtained from a

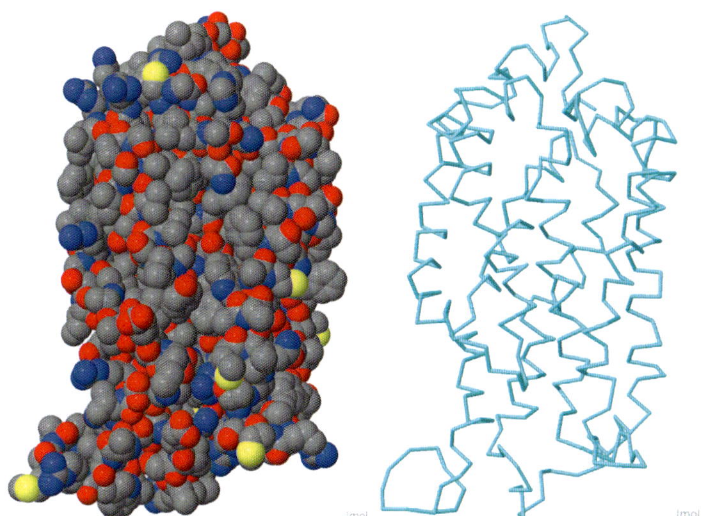


**Figure 6.34** Sphere (left panel a) and backbone (right panel b) representations of native human OR 17–40.

homology modeling as detailed in reference [Alfinito et al. (2009c)]. The template PDB entries were BR 1JFP and 1LN6 for the native and activated states, respectively, complemented with 1F88, chain A [Berman et al. (2000)].

Figures 6.34–6.35 report the sphere (left panel a) and backbone (right panel b) representations of the human OR 17–40 in its native and activated states, as obtained by the homology modeling procedure.

## 6.4.2 *Topological Properties*

The global insight on the protein conformational change, as induced by the ligand capture, is given by the contact maps reported in Fig. 6.36 and Fig. 6.37. To compare the native and activated states in the same figure, the native state is represented only with points $i < j$ while the activated state with points $i > j$. Furthermore, the contact maps are calculated with two different interaction radii, specifically: 6 Å in Fig. 6.36 and 20 Å in Fig. 6.37. From these figures one can observe that by increasing the value of $R_c$, the density of

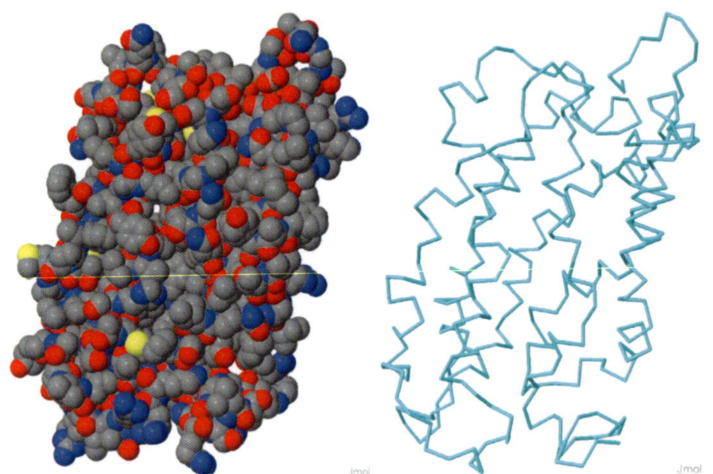


**Figure 6.35** Sphere (left panel a) and backbone (right panel b) representations of active human OR 17-40.

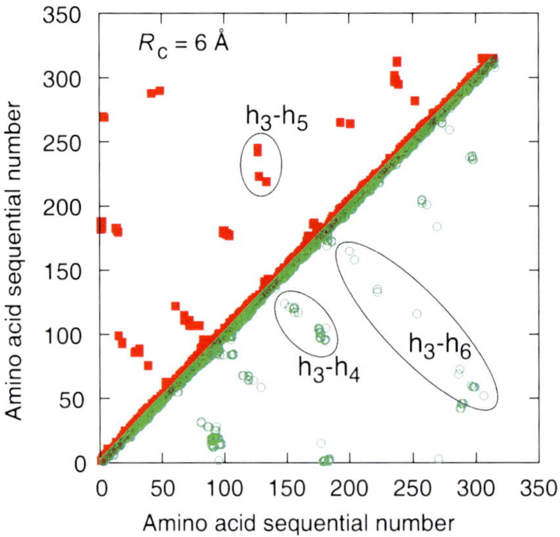


**Figure 6.36** Contact maps of human OR 17-40 for $R_c = 6$ Å. Results of the native state are reported on the left side of the diagonal (red full squares), and results of the activated state are reported on the right side of the diagonal (green open circles). The ellipses signal the main differences between the native and activated states, which are attributed to helices h3-h4-h6.

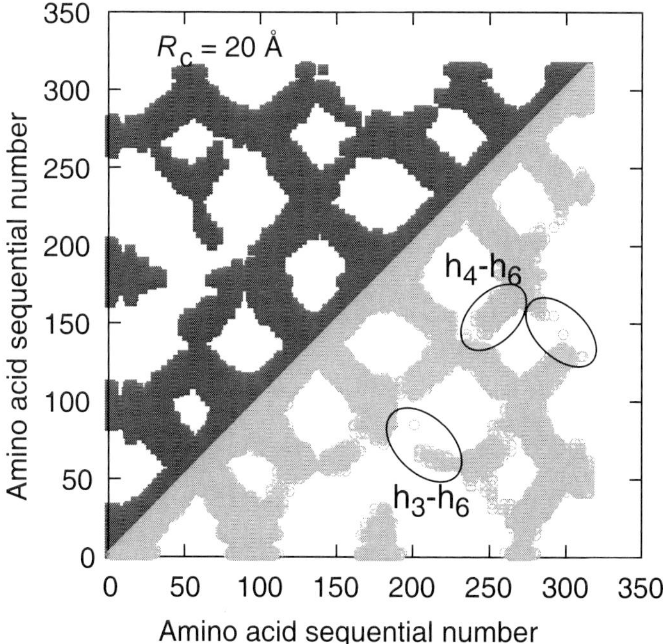


**Figure 6.37** Contact maps of human OR 17–40 for $R_c = 20$ Å. Results of the native state are reported on the left side of the diagonal (full squares), and results of the activated state are reported on the right side of the diagonal (open circles). The ellipses signal the main differences between the native and activated states, which are attributed to helices h3-h4-h6.

points and thus the connectivity increase substantially, as expected. In particular, the main differences between the native and activated states are in the increased connectivity among the helices h3-h4-h6, where the principal binding pocket is presumably located [Hall et al. (2004)]. Notably, significant differences between the native and activated states survive for large $R_c$, say up to $R_c$ values of about 50 Å.

## 6.4.3 *Protein Resistance*

The value of the protein resistance is in general different for the native and activated states, and the magnitude of this difference

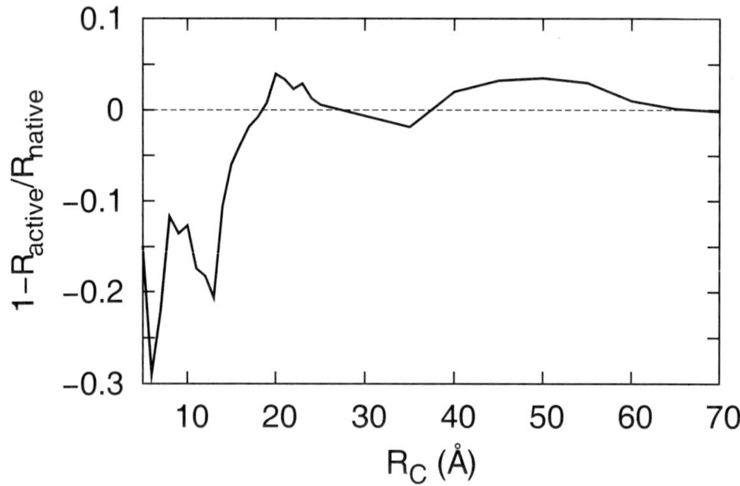


**Figure 6.38** RRV of human OR 17–40 as a function of the interacting radius $R_c$.

depends on $R_c$, as reported in Fig. 6.38. Here the main result is the nonmonotonic behavior of the resistance change at increasing values of $R_c$. The region of maximal sensitivity to the conformational change is found for $R_c$ in the range of 6–14 Å. Furthermore, in the ranges of $R_c$ values 18–26 Å and 38–65 Å, calculations give an evidence of an inversion of the resistance variation, with the activated state becoming less resistive than the native state. In terms of the electrical network, such an inversion is interpreted as a stronger increase of parallel with respect to series connections of the elementary resistances associated with the active links.

Recent advances in the activation mechanism of GPCRs [Kobilka (2007); Kobilka and Deupi (2007)] depict its dynamics not in terms of a simple on/off switch but, to some extent, in terms of a complex series of intermediate conformational transitions that involve the disruption of noncovalent intramolecular interactions. These interactions stabilize the protein state in a specific equilibrium condition, which depends on the ligand concentration. An increasing concentration of ligands induces the disruption of stabilizing interactions, enabling the receptor to evolve toward a more activated state.

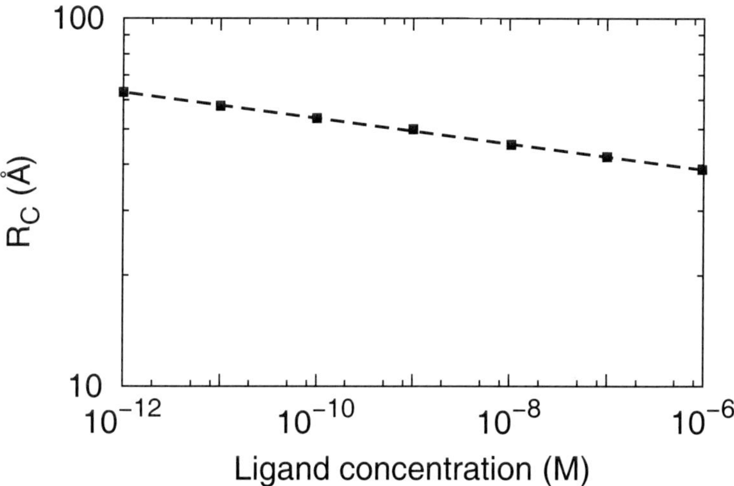


**Figure 6.39** Interacting radius $R_c$ as a function of the specific ligand concentration for human OR 17–40.

In this model, it is conjectured that the disruption of stabilizing interactions is related to a reduction in the interaction radius, which corresponds to a reduction in connections between amino acids. Thus, it is assumed that there exists a correlation between the interacting radius and the concentration of a specific odorant [Gether and Kobilka (1998); Kobilka (2007); Kobilka and Deupi (2007)]. In this way, the change in the single-protein resistance versus $R_c$ (see Fig. 6.38) is correlated with the analogue quantity measured for different ligand concentrations [Benilova et al. (2008a,b)]. To enforce this conjecture, Fig. 6.39 reports $R_c$ as a function of a specific odorant concentration. Here the data are obtained by comparing the experimental response for different helional concentrations [Benilova et al. (2008a,b)] with the theoretical RRV of the single protein at different $R_c$ values, as reported in Fig. 6.38. To this purpose, Fig. 6.40 shows the polarization resistance variation versus the odorant concentration as measured for the specific odorants heptanal and helional [Benilova et al. (2008a,b)]. Theoretical results (dashed curve) are found to reproduce qualitatively well the experiments (symbols), which provide evidence of a typical bell-shape behavior with maximum sensitivity for the case of helional at

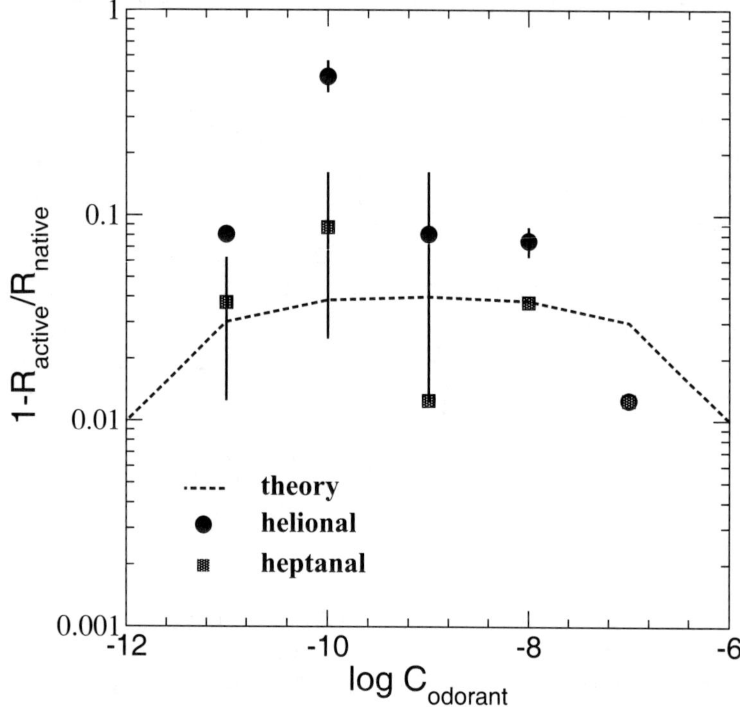


**Figure 6.40** Percent change in the polarization resistance at increasing concentration of specific odorants for human OR 17–40 at room temperature. Symbols refer to experiments with bars giving the estimated uncertainty [Benilova et al. (2008a,b); Hou et al. (2007)], and dashed line refers to theoretical calculations.

a molar concentration of $10^{-10}$ M. By construction, the quantitative comparison between theory and experiments cannot discriminate the different sensitivity of the OR to different odorants. In any case, at this stage of investigation, an overall agreement is found between theory and experiments, satisfactory enough to further support the proposed conjecture.

Interestingly enough, this kind of dose response departs from the saturation behavior observed for other receptors (like rat OR I7) where an increasing concentration of the specific ligand produces a systematic increase in the protein response [Hou et al. (2007)]. According to the previous conjecture, the origin of this difference

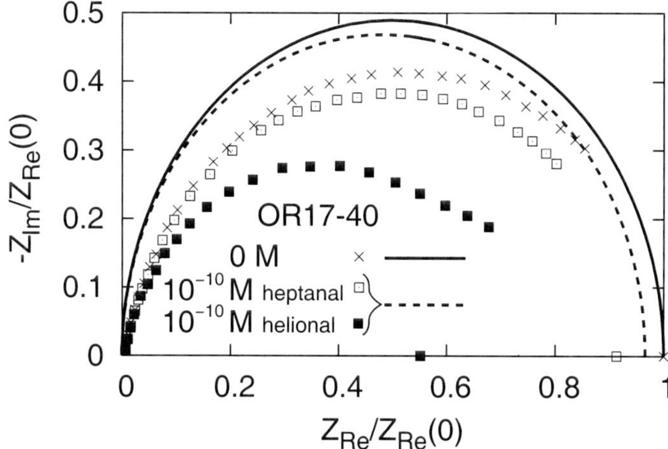


**Figure 6.41** Nyquist plot of human OR 17–40 in the absence and the presence of a specific ligand (heptanal, helional). The impedances are normalized to the static value of the native state, $Z_{Re}^{nat}(0) = 33$ KΩ cm$^2$. Symbols pertain to experiments with crosses referring to no odorant, and empty (full) squares to a heptanal (helional) concentration of $10^{-10}$ M at room temperature [Benilova et al. (2008a,b); Hou et al. (2007)]. Curves pertain to theoretical results with the continuous curve referring to the native state configuration as input data with $R_c = 70$ Å, and dashed line to the activated state configuration with $R_c = 46$ Å.

should be found in the peculiar topological modification undergone by human OR 17–40.

### 6.4.4 *Small-Signal Electrical Properties*

The impedance response of the single protein is explored over a wide range of frequencies, and the results are given by means of the Nyquist plot within a frequency range from 1 mHz to 100 kHz as in experiments [Benilova et al. (2008a,b)].

Figure 6.41 reports the Nyquist plots with the impedance normalized to the static value of the native state for human OR 17–40. Symbols pertain to experiments with crosses referring to the absence of a specific odorant, and empty (full) squares to the presence of the specific odorant heptanal (helional) at the concentrations reported in the figure. Curves pertain to theoretical results where the single protein is taken to be representative of the entire sample, and with continuous (dashed) lines referring to native (activated) state. Within the present model, Nyquist plots are obtained by using as input data the networks corresponding to the native and activated states at the cut-off radius that, according to Fig. 6.40, yields the maximum resolution. The agreement between theory and experiments is found to be qualitatively satisfactory and acceptable from a quantitative point of view. One can remark that the near-ideal semicircle shape of the experimental Nyquist plot is well reproduced, thus confirming that the network impedance model behaves closely to a single RC circuit as expected by the presence of a rather uniform distribution of time constants associated with the different values of the resistance and capacitance of the links [Alfinito et al. (2009c)]. The scarcity of experimental data and the lack of a certified knowledge for the 3D structures of the considered proteins lead one to consider these results as a first but significant step toward a microscopic modeling of the electrical properties of this olfactory receptor.

### 6.4.5 *Conclusion*

EIS experiments on human OR 17–40 showed that this OR can be used as an active element for a nanobiosensor [Benilova et al. (2008a,b)]. The mechanism of odorant capture is monitored by means of the modification of the impedance spectra, and these results are generally confirmed by surface plasmon resonance and bioluminescence response of appropriately prepared OR 17–40 samples [Benilova et al. (2008a,b); Marrakchi et al. (2007); Vidic et al. (2008, 2006)]. The EIS results are microscopically interpreted here on the basis of the conformational change in the protein tertiary structure induced by the sensing action. Accordingly, the conformational change induces a variation of the protein electrical response that is detectable with the EIS technique and can be reproduced by means of the INPA model. In particular, a possible correlation between the interacting radius, at the basis of charge transfer between amino acids, and the odorant concentration is inferred by comparing theory with experiments. From one side, at present the INPA approach is rather essential and the qualitative agreement with available experiments is used as a proof of concept that is promising for the realization of an electronic nose based on an array of nanobiosensors. From another side, relevant information on the protein tertiary structure and the mechanisms of charge transfer can be extracted from the microscopic interpretation of experiments.

## 6.5 OR 7D4

This section investigates the electrical and related properties of the single-sensing protein OR 7D4 in its structure belonging to human and chimpanzee. In particular, the change in the impedance spectrum following a conformational change is analyzed.

### 6.5.1 *Modeling*

In the absence of data from the PDB, the GPCR Automodeller online automatic procedure [Launay et al. (2012a,b)] is used for calculating the 3D structures of human and chimpanzee OR 7D4 in the native and activated states. To produce the native states, four reference proteins belonging to the class A GPCRs (1U19, 2RH1, 2VT4, and 3EML) have been considered. For each protein, a set of 10 templates, mainly different for the $C$–$N$ terminal tails, have been used for modeling the protein conformation. The activated state has a single reference protein, the type-2 opsin 3CAP.

Figures 6.42 and 6.43 report the sphere (left panel a) and backbone (right panel b) representations of the chimpanzee OR 7D4 in its native and activated states, as obtained by the homology modeling procedure.

### 6.5.2 *Topological Properties*

The global insight on the protein conformational change, as induced by the ligand capture, is given by the contact maps reported in Figs. 6.44 and 6.45.

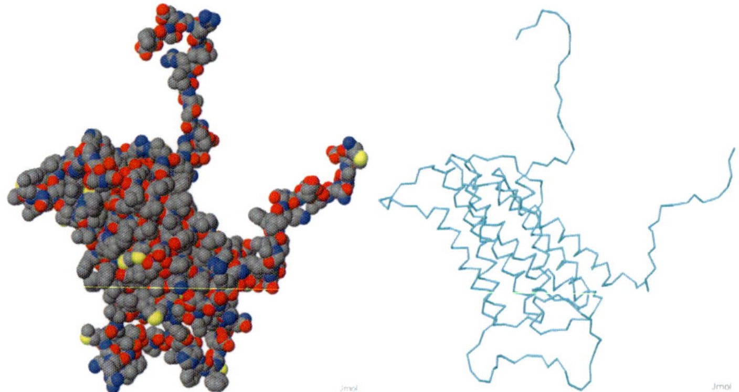


**Figure 6.42** Sphere (left panel a) and backbone (right panel b) representations of native chimpanzee OR 7D4.

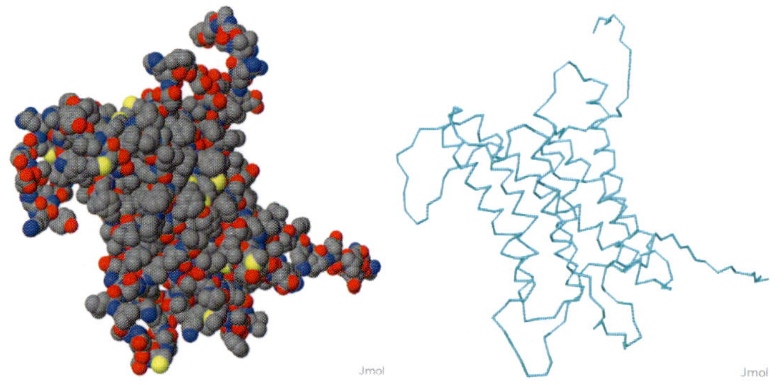


**Figure 6.43** Sphere (left panel a) and backbone (right panel b) representations of active chimpanzee OR 7D4.

## 6.5.3 *Protein Resistance*

By applying the INPA model, the RRV, that is, the normalized difference of resistance between the native and activated states, is calculated for different values of the interaction radius, $R_c$, for human and chimpanzee OR 7D4. In particular, by using all the templates, 400 different possible response profiles are produced. Of these, the 10 more representative are reported in the figures following. Furthermore, to explore the role of terminal tails in the

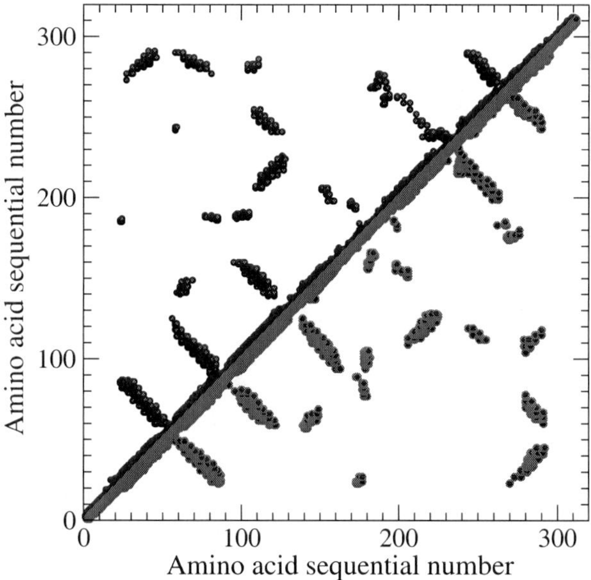


**Figure 6.44** Contact map of the two chimpanzee OR 7D4 representations: native state (circles on the left-hand side) and activated state (circles on the right-hand side). Because of the axial symmetry with respect to the diagonal exhibited by data, the symbols pertaining to the native map are reported only on the left-hand side of the diagonal and those pertaining to the activated map on the right-hand side. The interaction radius is $R_c = 10$ Å.

protein resistance responses, the 3D structures, deprived of the terminals, are also showed.

The main results are reported in Fig. 6.46 for the human and in Fig. 6.47 for the chimpanzee OR 7D4, and they can be summarized as follows. The main characteristics of the RRV are as follows:

(i) Different OR 7D4 couples of templates show similar profiles, with variations up to about 70% for the case of human OR 7D4 and up to about 150% for the case of chimpanzee OR 7D4.
(ii) In the presence/absence of the $N$–$C$ terminals, the RRV shape changes, going from a two-bump feature to a single-bump feature (see Figs. 6.47 and 6.48). These results suggest that the region less influenced by the presence of C and N terminals is that for $R_c \leq 20$ Å.

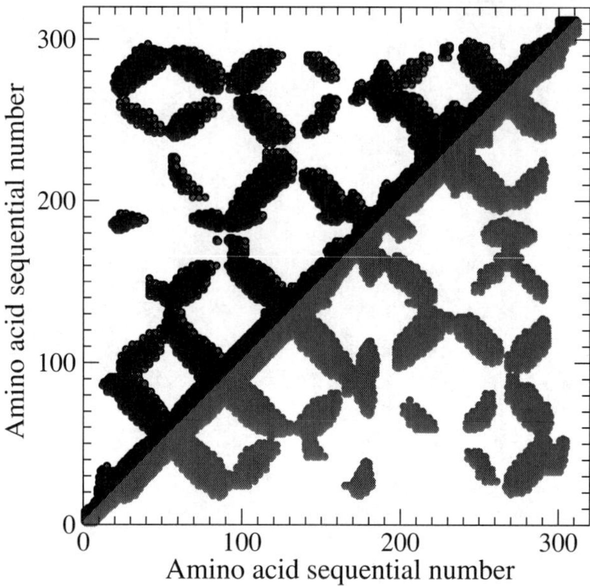


**Figure 6.45** The same as in Fig. 6.44 for an interaction radius of $R_c = 20$ Å.

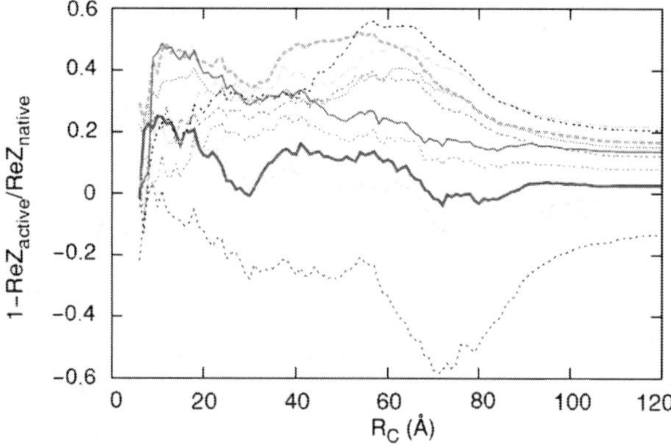


**Figure 6.46** Human OR 7D4: RRV versus $R_c$ for the set of 10 templates obtained with the native state of A2A human adenosine receptor 3EMLa (22% accuracy) and one template of the activated state of opsin 3CAPa. Data refer to the complete protein.

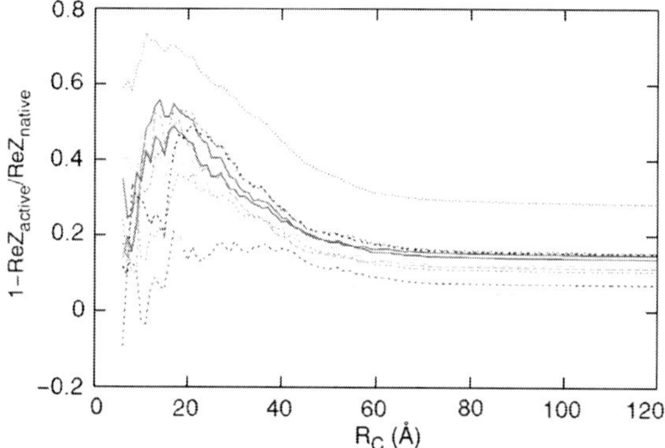


**Figure 6.47** Human OR 7D4: RRV versus $R_c$ for the set of 10 templates obtained with the native state of A2A human adenosine receptor 3EMLa (22% accuracy) and one template of the activated state of opsin 3CAPa. Data refer to the protein without the *N* and *C* terminals.

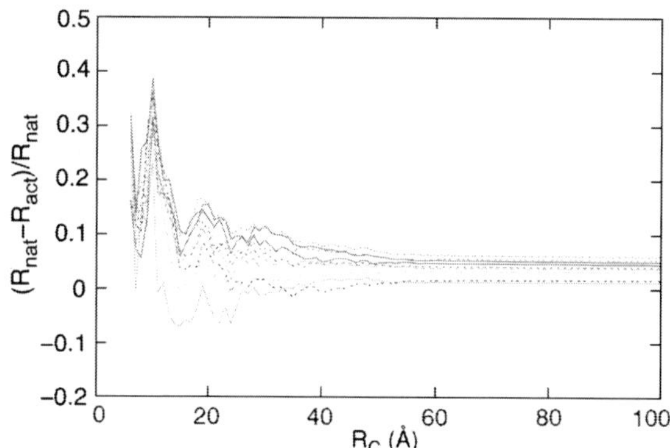


**Figure 6.48** Chimpanzee OR 7D4: RRV versus $R_c$ for the set of 10 templates obtained with the native state of A2A human adenosine receptor 3EMLa (22% accuracy) and one template of the activated state of opsin 3CAPa. Data refer to the protein without the *N* and *C* terminals.

### 6.5.4 *Small-Signal Electrical Properties*

The impedance frequency spectrum of the chimpanzee OR 7D4 is reported as Nyquist plot, respectively, in Figs. 6.50 and 6.51 for two values of the interaction radius $R_c$ and for a complete protein or for a protein without *N* and *C* terminals. Both spectra

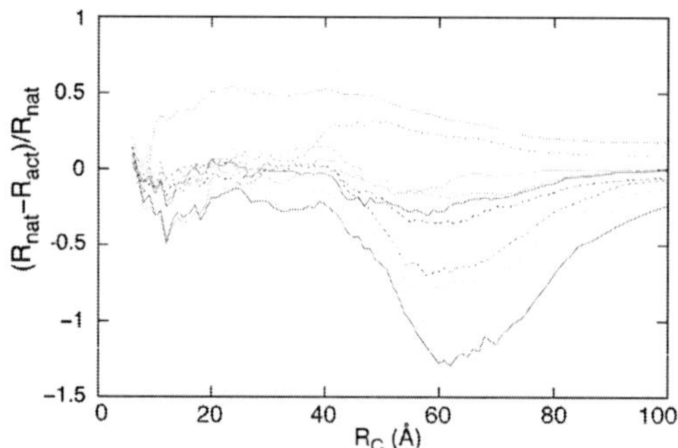


**Figure 6.49** Chimpanzee OR 7D4: RRV versus $R_c$ for the set of 10 templates obtained with the native state of A2A human adenosine receptor 3EMLa (22% accuracy) and one template of the activated state of opsin 3CAPa. Data refer to the complete protein that includes *N* and *C* terminals.

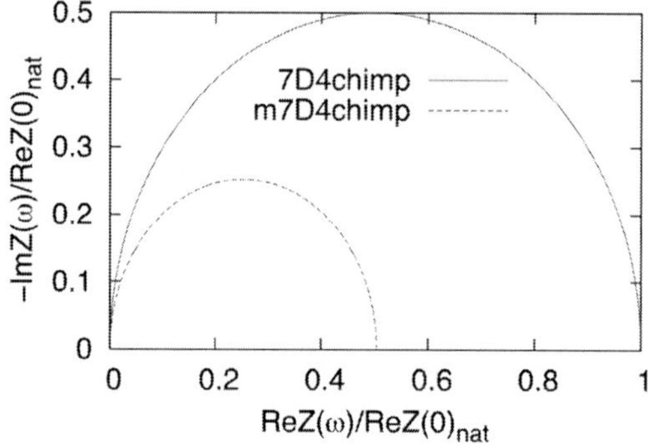


**Figure 6.50** Nyquist plot of the normalized impedance of chimpanzee OR 7D4. Calculations are carried out with the native state providing the maximum RRV using A2A human adenosine receptor 3EMLa (22% accuracy) and one template of the activated state of opsin 3CAPa. Data refer to the complete protein.

provide evidence of a near-ideal semicircle shape, thus confirming the presence of a uniform set of relaxation times. Furthermore, the presence of *N* and *C* terminals leads to a larger sensitivity of the complete protein with respect to the protein without terminals.

Preliminary EIS experiments performed on chimpanzee OR 7D4 at the University Claude Bernard I in Lyon (Jaffrezic et al., private communication), in the absence and presence of the specific

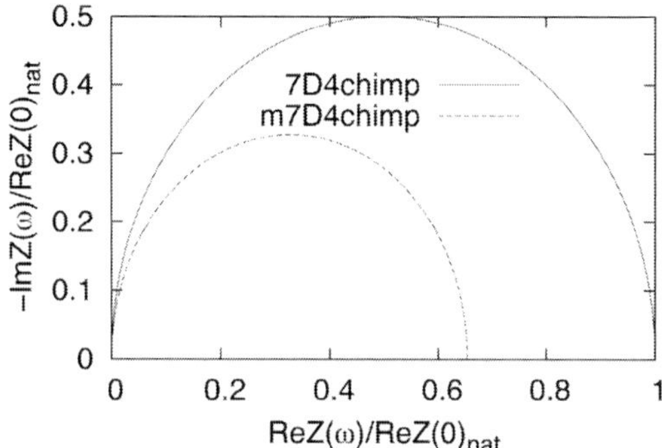


**Figure 6.51** Nyquist plot of the normalized impedance of chimpanzee OR 7D4. Calculations are carried out with the native state providing the maximum RRV using A2A human adenosine receptor 3EMLa (22% accuracy) and one template of the activated state of opsin 3CAPa. Data refer to the protein without the $N$ and $C$ terminals.

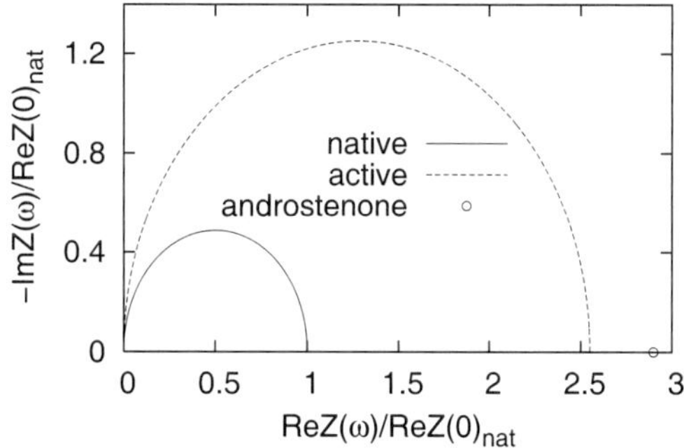


**Figure 6.52** Nyquist plots for the native and activated states of the selected 3D structures of chimpanzee OR 7D4. The interaction radius $R_c = 62$ Å gives the maximal resolution.

odorant androstenone, provide evidence of an increase for about a factor of three of the sample polarization resistance with respect to the native state as reported in Fig. 6.52 together with the theoretical predictions of the Nyquist plot for the most favorable protein structures. The experimental Nyquist plots at different concentrations of the odorant show a rather ideal semicircle shape; their fitting, performed by Zplot software, provides the change in the polarization resistance at increasing odor concentration as reported

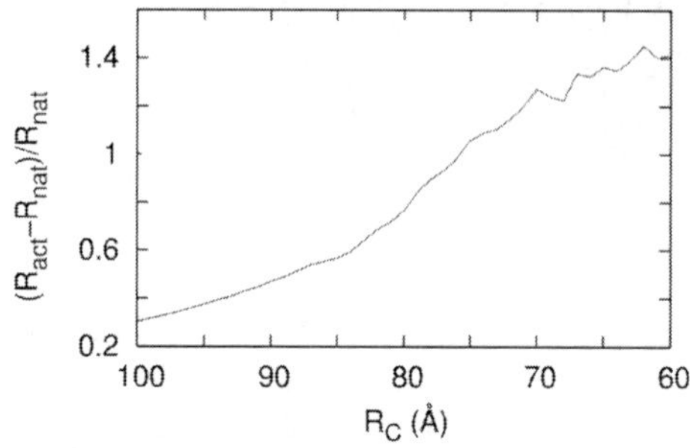


**Figure 6.53** Predicted maximum RRV for the complete chimpanzee OR 7D4. Decreasing values of $R_c$ are correlated to an increasing concentration of the specific odorant.

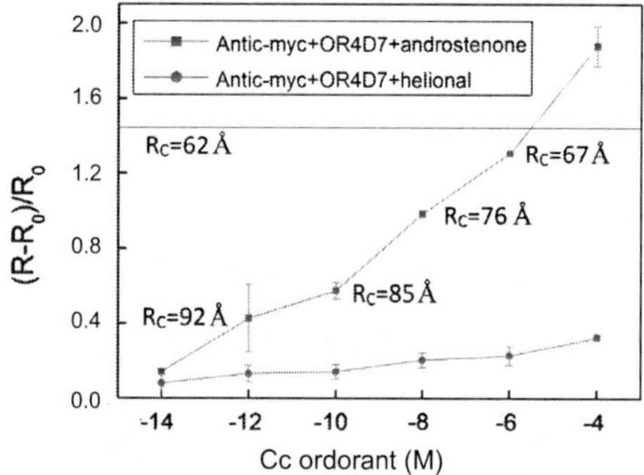


**Figure 6.54** Experiments versus predicted data calculated from RRV versus $R_c$ for the complete chimpanzee OR 7D4 as reported in Fig. 6.53.

in Fig. 6.54. Here a systematic increase in the normalized change in resistance at increasing odorant concentration is compared with a maximum variation of about 300% at the highest concentration. The specific sensitivity was tested with success by substituting androstenone with helional, as reported in the same figure by the curve with circles.

The microscopic interpretation of these results is carried out by selecting among the possible 3D structures of the single protein (native/activated state) those that best approximate the experiments. Figure 6.54 summarizes a sampling of possible results from which the structures associated with the curve with the minimum value at $R_c = 62$ Å are considered. The corresponding theoretical Nyquist plots of the native and activated states are calculated and the results are reported in Fig. 6.52. The agreement between theory and experiments is found to be satisfactory both from a qualitative and quantitative point of view.

### 6.5.5 *Conclusion*

The conjecture concerning the possibility to describe the different protein responses to different odorant concentration by means of different choices of the cut-off value of the INPA is favorably tested also in this case, as shown in Fig. 6.54. Here the $R_c$ values corresponding to the experimental result are reported in the same figure. The fit is found to be possible up to $R_c = 67$ Å.

## 6.6 Human OR 2AG1

This section reports the electrical and related properties of the single-sensing protein human OR 2AG1. In particular, the change in the impedance spectrum following a conformational change is investigated.

### 6.6.1 *Modeling*

In the absence of data from the PDB, the online automatic procedure GPCR Automodeller [Launay et al. (2012a,b)] is used for calculating the 3D structures of human OR 2AG1 in the native and activated states. For both states, a set of 10 templates was available.

Figures 6.55 and 6.56 report the sphere (left panel a) and backbone (right panel b) representations of the human OR 2AG1 in its native and activated states, as obtained by the above procedure.

### 6.6.2 *Topological Properties*

The global insight on the protein conformational change, as induced by the ligand capture, is given by the contact maps reported in Figs. 6.57 and 6.58 for two values of the interaction radius. In these figures, the data reproduce the couples of connected amino acids for the template number 7 of the native and activated states, with two

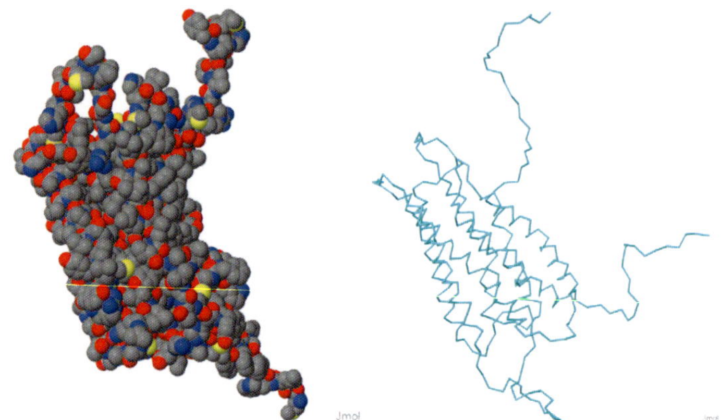


**Figure 6.55** Sphere (left panel a) and backbone (right panel b) representations of the native human OR 2AG1.

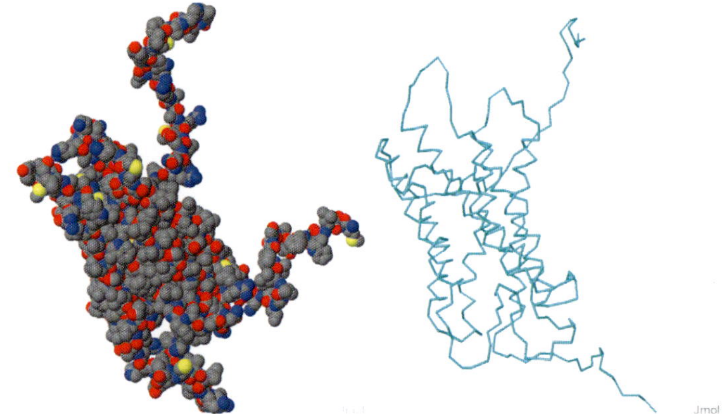


**Figure 6.56** Sphere (left panel a) and backbone (right panel b) representations of the active human OR 2AG1.

different values of $R_c = 9$ Å and 40 Å, respectively. In both the cases, the differences between the native and activated states are relevant and cover the whole structure.

### 6.6.3 *Protein Resistance*

By applying the INPA model, the RRV is calculated for different values of the interaction radius, $R_c$. Homonym templates, related

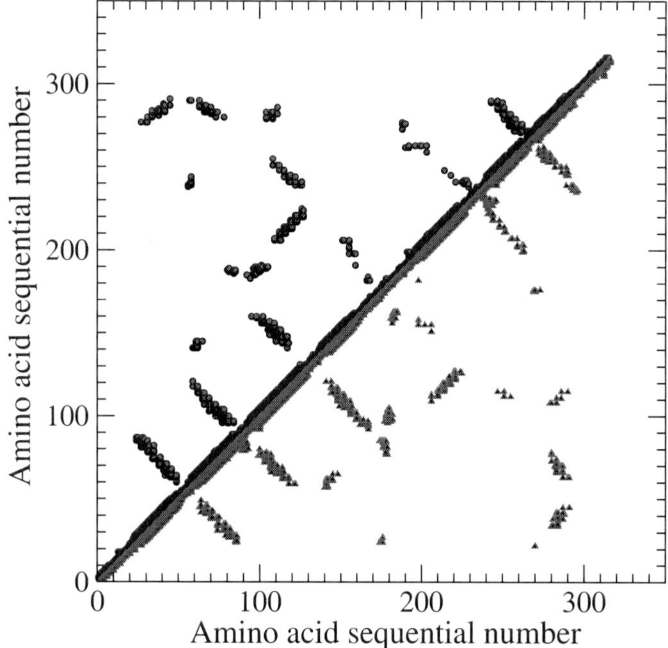


**Figure 6.57** Contact map of the two human OR 2AG1 representations: native state (circles on the left) and activated state (triangles on the right). Because of the axial symmetry with respect to the diagonal exhibited by data, the symbols pertaining to the native state are reported only on the left-hand side of the diagonal and those pertaining to the activated state on the right-hand side. The interaction radius is $R_c = 9$ Å.

to the native and activated states, have been used in calculations, and the spectra of 10 different structures have been obtained. Among them, the data referring to the template number 7 are reported here as those best reproducing the experimental data. The value of the protein resistance is in general different for the native and activated states, and the magnitude of this difference depends on $R_c$, as reported in Fig. 6.59. Here the main result is the nonmonotonic behavior of the resistance change at increasing values of $R_c$. Furthermore, it can be noticed that a fixed amount of difference is present also for large values of $R_c$. This is compatible with a residual basal activity [Gether and Kobilka (1998); Kobilka and Deupi (2007)].

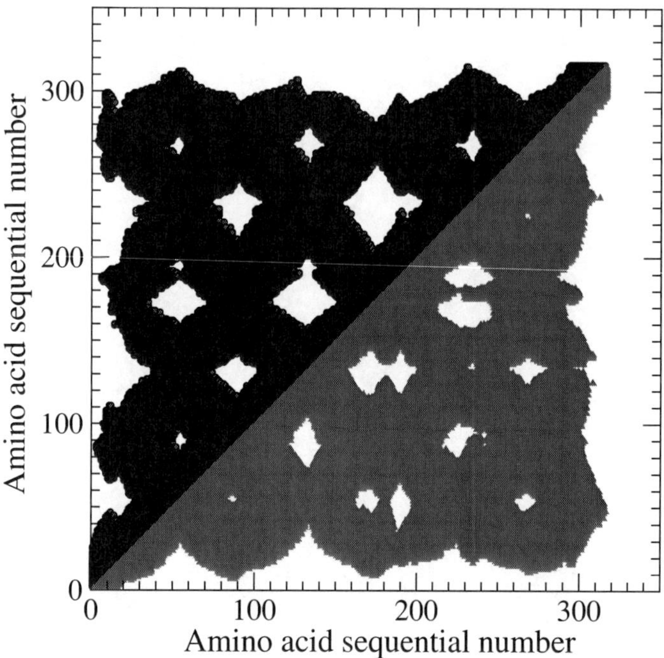


**Figure 6.58** The same as in Fig. 6.57 for an interaction radius $R_c = 40$ Å.

Besides the RRV, also the percentage variation of conductance is calculated and reported in Fig. 6.60. Here the variation of conductance in going from the native to the activated state for the template number 7 exhibits a minimum at about $R_c = 60$ Å which is common to all the considered templates.

The numerical results in Fig. 6.60 are then compared with the experimental data obtained by using single-walled carbon nanotubes (SWCNT) functionalized to anchor this protein as reported in Fig. 3b of Lee et al. (2012). In this case, the large minimum around $R_c =$ 60 Å should correspond to the reference value taken for a negligible concentration of the odorant amyl butyrate. By decreasing the values of the interaction radius, in agreement with the experimental data, the percentage conductance variation increases on the basis of a correlation between the increase in odorant concentration and the decrease in interacting radius.

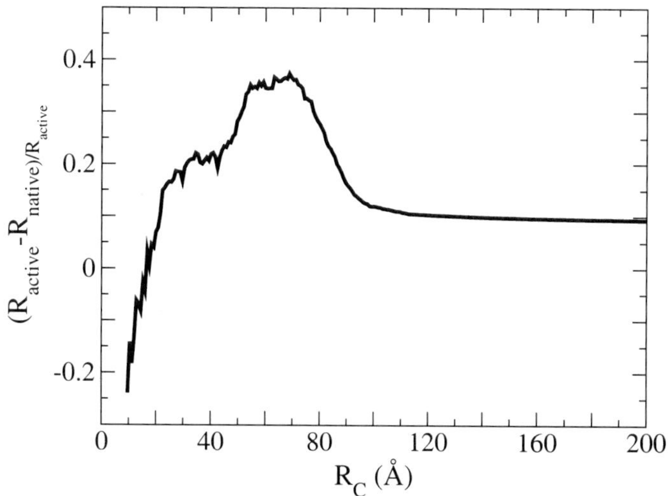


**Figure 6.59** RRV of human OR 2AG1, template number 7, as a function of the interacting radius $R_c$.

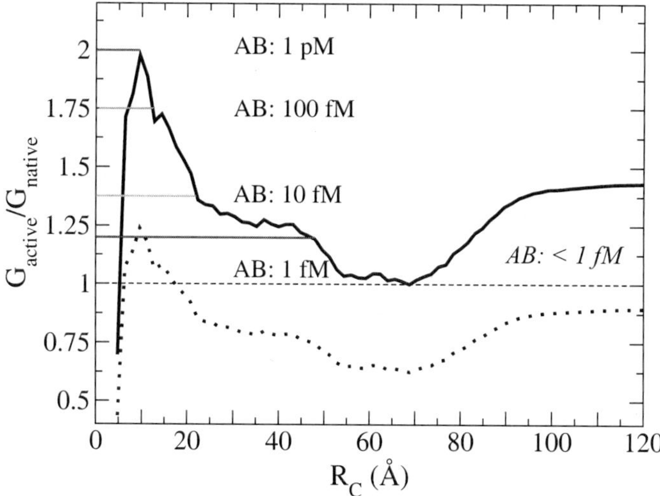


**Figure 6.60** Normalized conductance variation of human OR 2AG1, template number 7, as a function of the interacting radius $R_c$. The dashed curve refers to the numerical results, and the continuous curve refers to the values rescaled to give one at $R_c = 68$ Å. The lines give the experimental variations of conductance at the corresponding concentration of the amyl-butyrate odorant.

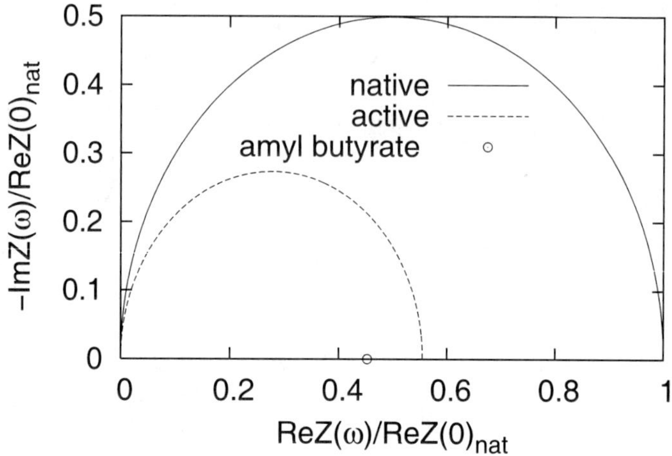


**Figure 6.61** Nyquist plot of human OR 2AG1. Curves pertain to theoretical results obtained with the template number 7: continuous line refers to the native state configuration as input data, and dashed line refers to the activated state with $R_c = 9$ Å.

## 6.6.4 *Small-Signal Electrical Properties*

The impedance response of the single-protein human OR 2AG1 is analyzed over a wide range of frequencies, and the results are given by means of the Nyquist plot as reported in Fig. 6.61 with the impedance normalized to the static value of the native state. Curves pertain to theoretical results where the single protein is taken to be representative of the entire sample, and with continuous (dashed) lines referring to native (activated) state. The choice of $R_c$ is taken as that giving the highest conductance variation (see Fig. 6.60) corresponding to the injection of 1 pM of AB to the SWCNT device [Park et al. (2012)]. The agreement between theory and experiments is found to be qualitatively satisfactory and acceptable from a quantitative point of view. One should remark the deviation from a near-ideal semicircle shape of the theoretical Nyquist plot, which is expected for small values of the interaction radius [Alfinito et al. (2009c)]. This simply means that the time constants associated with the different values of the resistance and capacitance of the links are not uniformly distributed. In particular, in the present case, two main time constants are prevalent.

### 6.6.5 *Conclusion*

Recent I–V experiments have developed a new technology for the immobilization of human OR 2AG1 onto SWCNT devices, working like field-effect transistors (FETs) [Lee and Park (2010); Lee et al. (2012)]. The main result of these studies is that this technology is very effective for producing electronic noses based on olfactory proteins. In particular, by monitoring the conductance variation of these devices, concentrations of the specific odorant, AB, can be detected below about 1 pM. The small-signal spectra are microscopically interpreted here on the basis of the conformational change in the protein tertiary structure induced by the sensing action. Accordingly, the conformational change induces a variation of the protein electrical response that is detectable with the conductance measurements and can be reproduced by means of the INPA model. In particular, a possible correlation between the interacting radius at the basis of charge transfer between amino acids and the odorant concentration is inferred by comparing theory with experiments [Alfinito et al. (2011c)].

## 6.7 Canine Cf OR 5269

This section reports on the electrical and related properties of the single-sensing protein canine Cf OR 5269. In particular, the change in the impedance spectrum following a conformational change is investigated.

### 6.7.1 *Modeling*

In the absence of data from the PDB, the online automatic procedure GPCR Automodeller [Launay et al. (2012a,b)] is used for calculating the 3D structures of canine Cf OR 5269 in the native and activated states.

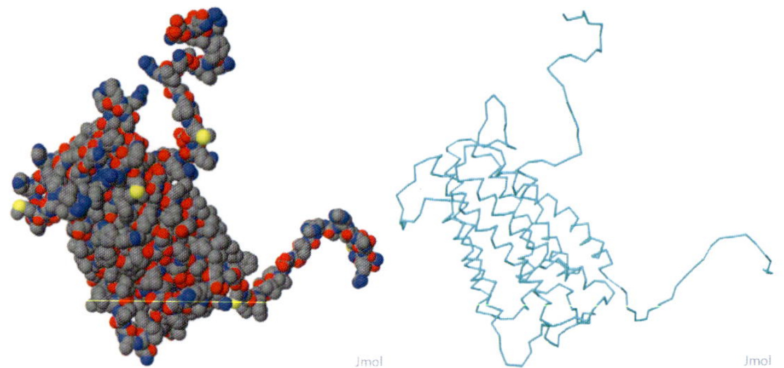


**Figure 6.62** Sphere (left panel a) and backbone (right panel b) representations of the native canine Cf OR 5269.

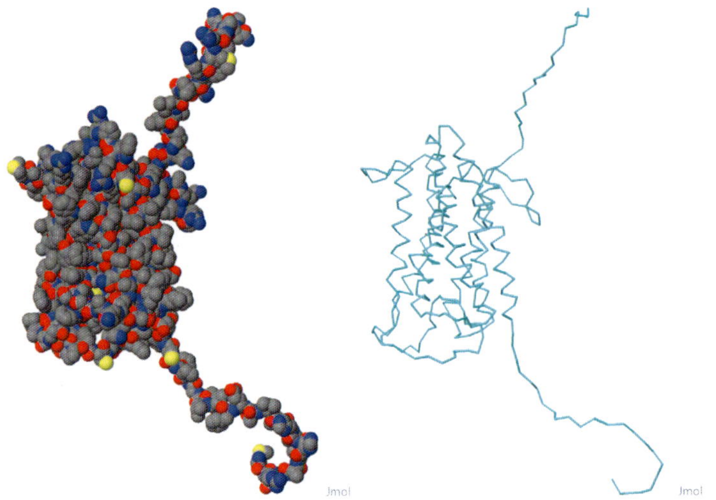


**Figure 6.63** Sphere (left panel a) and backbone (right panel b) representations of the activated canine Cf OR 5269.

Figures 6.62–6.63 report the sphere (left panel a) and backbone (right panel b) representations of canine Cf OR 5269 in its native and activated states, as obtained by the homology modeling procedure.

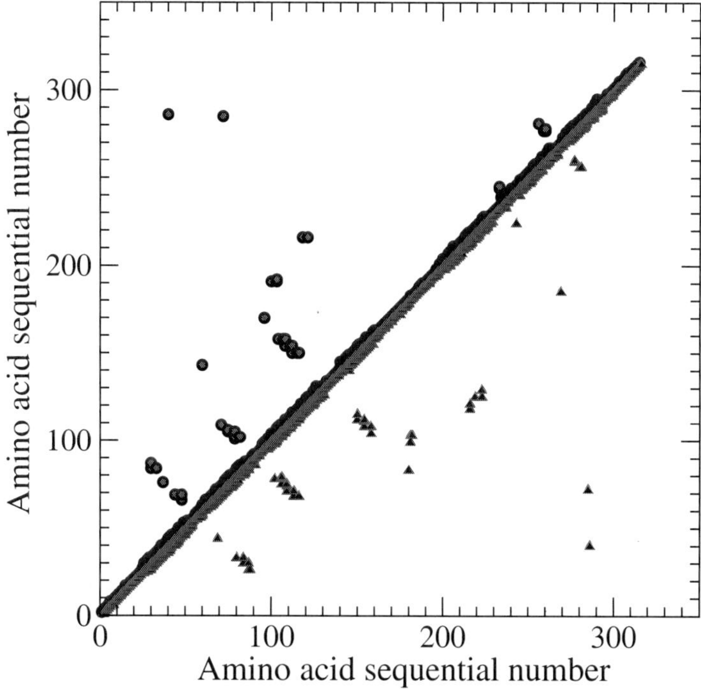


**Figure 6.64** Contact map of the two canine Cf OR 5269 representations: native state (circles on the left) and activated state (triangles on the right). Because of the axial symmetry with respect to the diagonal exhibited by data, the symbols pertaining to the native map are reported only on the left-hand side of the diagonal and those pertaining to the activated map on the right-hand side. The interaction radius is $R_c = 6$ Å.

## 6.7.2 *Topological Properties*

The global insight on the protein conformational change, as induced by the ligand capture, is given by the contact maps reported in Fig. 6.64 and Fig. 6.65 for two values of the interaction radius. In these figures, each couple of connected amino acids, say $i$, $j$, is represented with a point of coordinates $i$, $j$. The diagonal is the symmetry axis. To compare the native and activated states in the same figure, the native state is represented only with points $i < j$ while the activated state with points $i > j$.

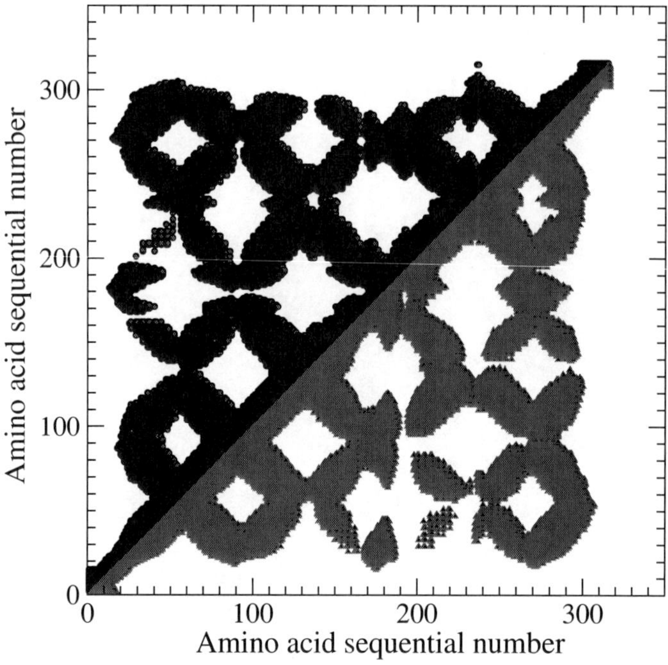


**Figure 6.65** The same as in Fig. 6.64 for an interaction radius of $R_c = 27$ Å.

### 6.7.3 *Protein Resistance*

By applying the INPA model, the RRV is predicted at different values of the interaction radius, $R_c$. Homonym templates related to the native and activated states have been used for the calculations; in this way the spectrum of 10 different sets is obtained. Among them the data referring to the template number 3 are reported here as those best reproducing the experimental data [Park et al. (2012)]. The value of the protein resistance is in general different for the native and activated states, and the magnitude of this difference depends on $R_c$, as reported in Fig. 6.66. Here the main result is the nonmonotonic behavior of the resistance change at increasing values of $R_c$. The region of maximum sensitivity to the conformational change is found for $R_c$ around 20 and 60 Å, respectively, with the resistance of the activated state being larger than that of the native state.

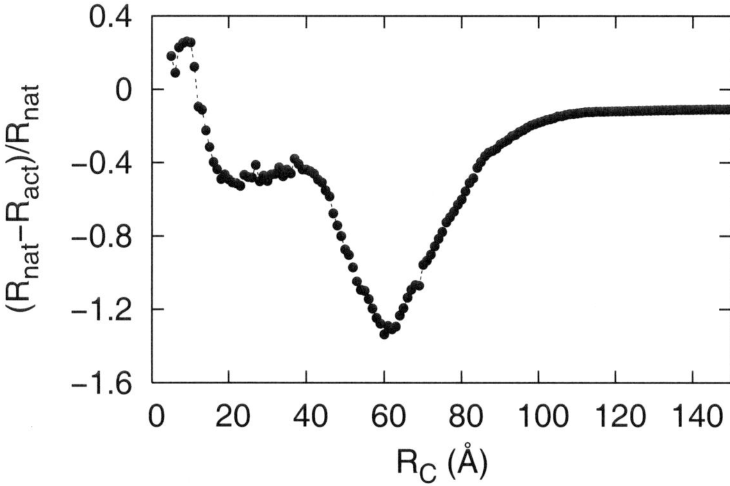


**Figure 6.66** RRV of canine Cf OR 5269 as a function of the interacting radius $R_c$.

### 6.7.4 *Small-Signal Electrical Properties*

The impedance response of the single protein is explored over the usual wide range of frequencies and the results are given by means of the Nyquist plot. Figure 6.67 reports the Nyquist plots with the impedance normalized to the static value of the native state for canine Cf OR 5269. Curves pertain to theoretical results where the single protein is taken to be representative of the entire sample, and with continuous and dashed lines referring to the native and activated states, respectively. Within the INPA model, the Nyquist plots are obtained by using as input data the networks corresponding to the native and activated states at the cut-off radius, which, according to Fig. 6.66, gives the maximum resolution. The agreement between theory and experiments is found to be qualitatively satisfactory and acceptable from a quantitative point of view. One should remark the near-ideal semicircle shape of the theoretical Nyquist plot, thus confirming that the network impedance model behaves closely to a single RC circuit as expected by the presence of a rather uniform distribution of time constants associated with the different values of the resistance and capacitance of the links [Alfinito et al. (2009c)].

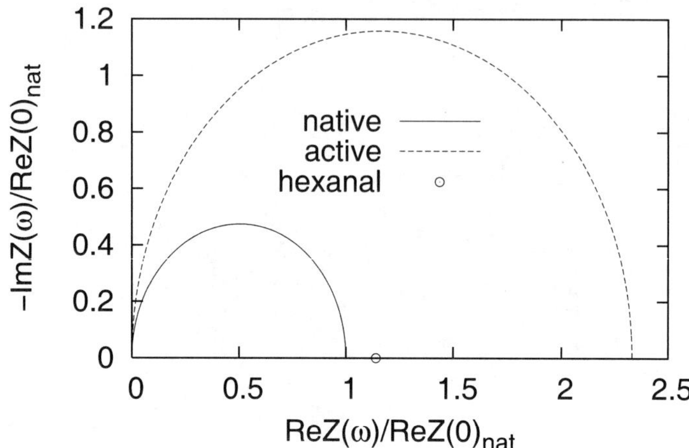


**Figure 6.67** Nyquist plot of canine Cf OR 5269 in the absence and in the presence of a specific ligand. The impedances are normalized to the static value of the native state, $Z^{nat}_{Re}(0) = 33\,\mathrm{K}\,\Omega\,\mathrm{cm}^2$. Empty circle refers to the experimental datum with an external concentration of $10^{-10}$ M at room temperature [Benilova et al. (2008a,b)]. Curves pertain to theoretical results with continuous curve referring to the native state configuration as input data with $R_c = 60$ Å and dashed line referring to the activated state configuration with $R_c = 60$ Å.

## 6.7.5 *Conclusion*

Measurements of change in carbon nanotube transistor conductance on which canine Cf OR 5269 was deposited showed that this OR can be used as an active element for a nanobiosensor [Park et al. (2012)]. The mechanism of odorant capture is monitored by means of the modification of the transistor conductance at increasing concentration of the specific odorant. The small-signal spectra are microscopically interpreted here on the basis of the conformational change of the protein tertiary structure induced by the sensing action. Accordingly, the conformational change induces a variation of the protein electrical response that is detectable with the conductance measurements and can be reproduced by means of the INPA model. In particular, a possible correlation between the interacting radius at the basis of charge transfer between amino acids and the odorant concentration is inferred by comparing theory with experiments [Alfinito et al. (2011c)].

## 6.8 Azurin

The aim of this section is to report a microscopic interpretation of the current–voltage characteristics of azurin, a blue-copper metallic protein that recently attracted large interest for possible applications in bioelectronics. Azurin is a metalloprotein, i.e., a protein complexed with an ionic metal, specifically copper. Copper may exist in the reduced state Cu(I) (here taken as the native state of azurin) and the oxidized state Cu(II) (here taken as the activated state of azurin); in the latter case, an electron is made free to transport energy.

### 6.8.1 *Modeling*

From the PDB, the 1AZU structure is taken, which is the structural feature of azurin by *Pseudomonas aeruginosa* at 2.7 Å of resolution.

Figure 6.68 shows the ribbon-like representation of azurin biological molecule. Figures 6.69–6.70 report the sphere (left panel a) and backbone (right panel b) representations of azurin in its native and activated states, as obtained by the homology modeling procedure.

### 6.8.2 *Topological Properties*

The global insight on the protein conformational change, as induced by the ligand capture, is given by the contact maps reported in Fig. 6.71 and Fig. 6.72 for two values of the interaction radius. The investigation on the protein topological properties has been performed by using the reduced and oxidized states of the protein variant II found in *Alcaligenes xylosoxidans*. The structures are present in the PDB and named, respectively, 1DZ0 and 1DYZ. The corresponding contact maps have been calculated for $R_c = 6$ Å and 20 Å, respectively. For a given interaction radius, significant differences between structures are not detectable.

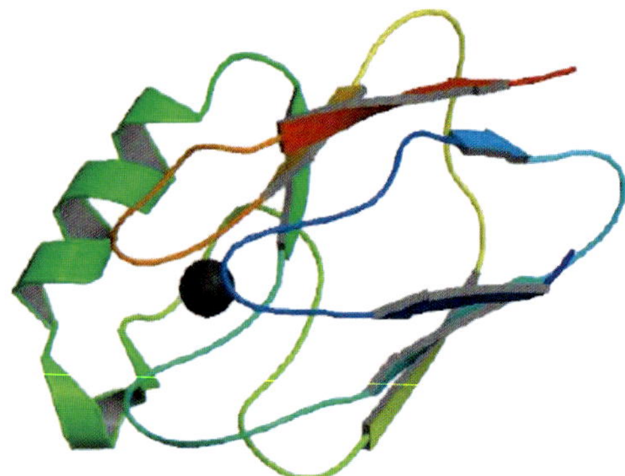

**Figure 6.68** Ribbon-like structure of azurin biological molecule (Image courtesy: Torsten Schwede, Konstantin Arnold/Swiss-Model Template Library).

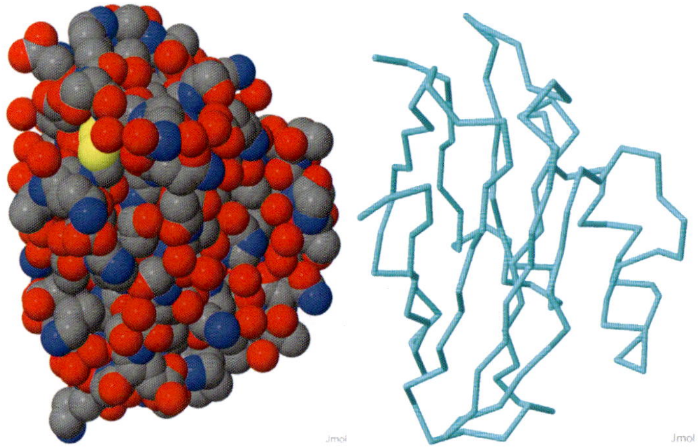


**Figure 6.69** Sphere (left panel a) and backbone (right panel b) representations of the native azurin 1DYZ.

## 6.8.3 *Protein Resistance*

By applying the INPA model, the RRV is predicted at different values of the interaction radius, $R_C$, and the reduced and oxidized forms of azurin are compared. The value of the protein resistance is in general very similar for both states, in agreement with the structural investigations that do not reveal significative differences in the

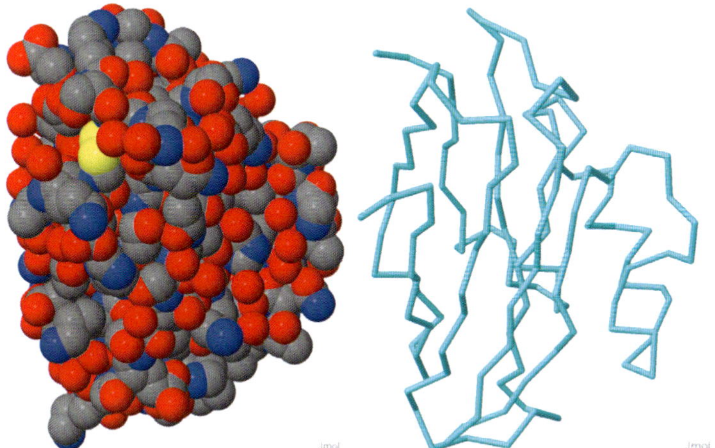


**Figure 6.70** Sphere (left panel a) and backbone (right panel b) representations of the active azurin 1DZ0.

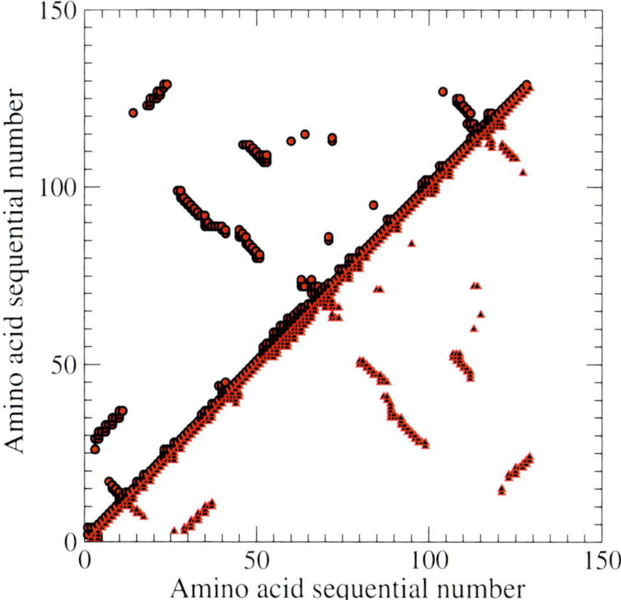


**Figure 6.71** Contact maps of azurin from *Alcaligenes xylosoxidans* for $R_c$ = 6 Å. Data of the reduced state are reported on the left side of the diagonal (circles), and data of the oxidized (activated) state are reported on the right side of the diagonal (triangles).

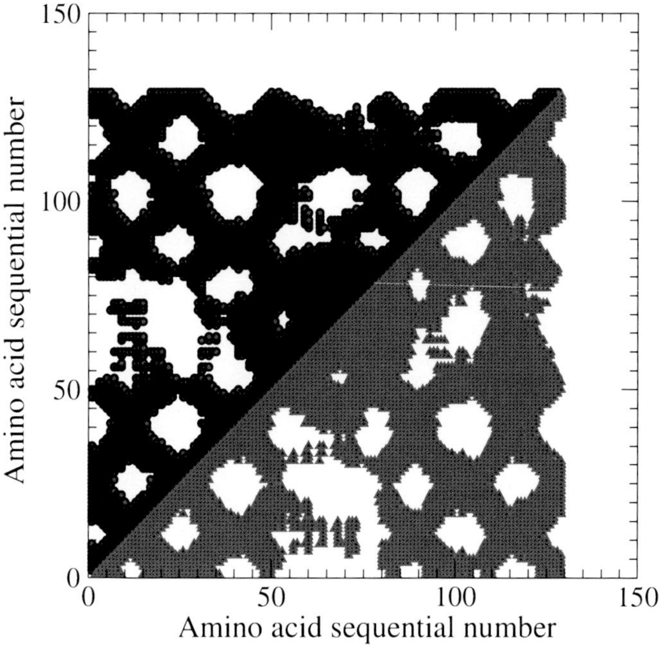


**Figure 6.72** Contact maps of azurin from *Alcaligenes xylosoxidans* for $R_c$ = 20 Å. Data of the reduced state are reported on the left side of the diagonal (circles), and data of the oxidized state are reported on the right side of the diagonal (triangles).

protein structures for different states of the Cu atom [Dodd et al. (2000)]. Calculations of the RRV are reported in Fig. 6.73.

### 6.8.4 *Current–Voltage Characteristics*

The current–voltage characteristics of a nanolayer of azurin deposited on an SH-terminated Si substrate and measured with the AFM technique are reported in Fig. 6.74 [Ron et al. (2010)]. Here theoretical calculations carried out on a single protein using an interacting radius $R_c = 6$ Å, a $\rho_{max} = 4 \times 10^8$ ΩÅ $\rho_{min} = 4 \times 10^5$ ΩÅ, and a barrier height of 0.219 eV are reported as black crosses in the same figure. The agreement between theory and experiments is found to be satisfactory.

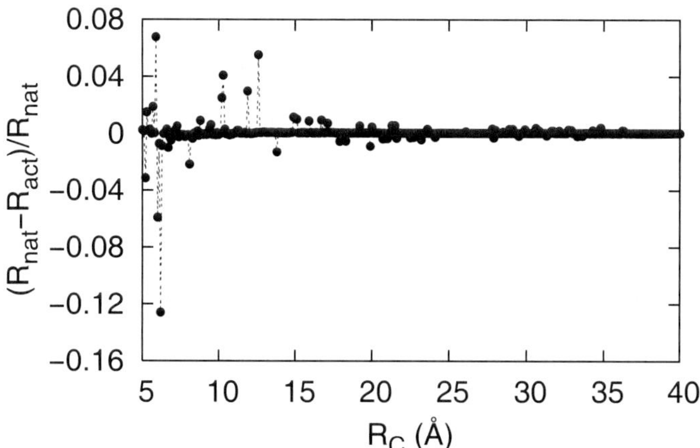


**Figure 6.73** RRV of azurin from *Alcaligenes xylosoxidans*, as a function of the interacting radius $R_c$.

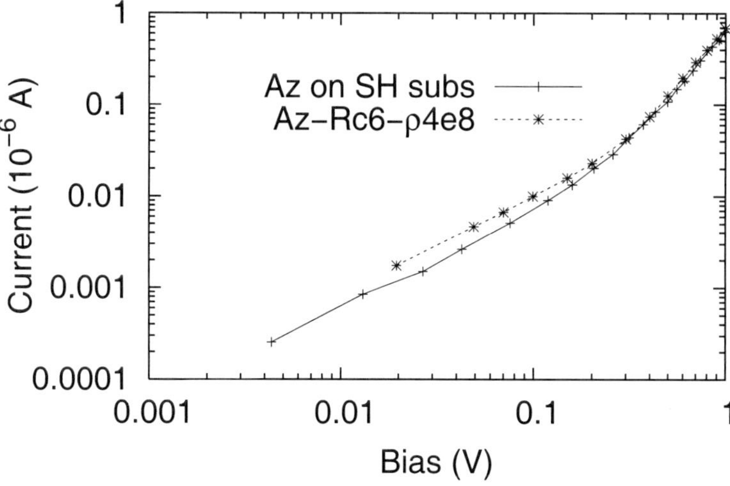


**Figure 6.74** Current–voltage characteristics of azurin. Continuous curve and crosses refer to experiments of azurin on SH substrate from reference [Ron et al. (2010)], and dashed curves and asterisk refer to theoretical simulations using an interacting radius $R_c = 6$ Å, a $\rho_{max} = 4 \times 10^8$ ΩÅ and a barrier height of 0.219 eV.

### 6.8.5 *Conclusion*

The metalloprotein azurin shows quite interesting electrical properties when used in a metal–protein–metal device [Ron et al. (2010)]. It exhibits an I–V characteristic very similar to that shown by the bR photoreceptor. The INPA model is able to reproduce the I–V characteristic of this protein by using the same set of parameters of bR, in particular, the same barrier height. Azurin obtained from the organism *Alcaligenes xylosoxidans* has also been analyzed in the reduced/oxidized form of the Cu atom, and the theoretical expectations agree with the structural information.

## 6.9 AChE

This section reports the electrical and related properties of the enzyme acetylcholinesterase (AChE) in the native state and in the complexed form (activated state) with Huperzine A. This protein plays a fundamental role in the process of functioning of muscle cells. It destroys the neurotransmitter acetylcholine after it has passed to these cells the information coming from the brain, so that new signals can be transmitted. Starting from the atomic coordinates, as reported in the PDB [Berman et al. (2000)], first the topological properties of the corresponding networks are investigated as a function of the interaction radius. Then the analogous investigation is carried out for the total impedance spectrum of the corresponding impedance network.

### 6.9.1 *Modeling*

Torpedo acetylcholinesterase is a globular protein, made of 14 $\beta$-sheets and 16 $\alpha$-helices. It is an enzyme that breaks the neurotransmitter acetylcholine into acetic acid and choline, thus stopping the transmission of signal from nerve cells to muscle cells.

At present, in the PDB [Berman et al. (2000)], there are 56 entries related to torpedo acetylcholinesterase. AChE is reported in many incomplete, different representations, either native or complexed

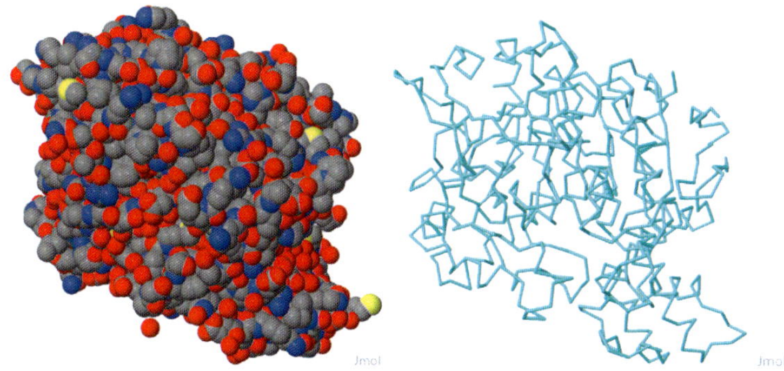


**Figure 6.75** Sphere (left panel a) and backbone (right panel b) representations of the native AChE 2ACE.

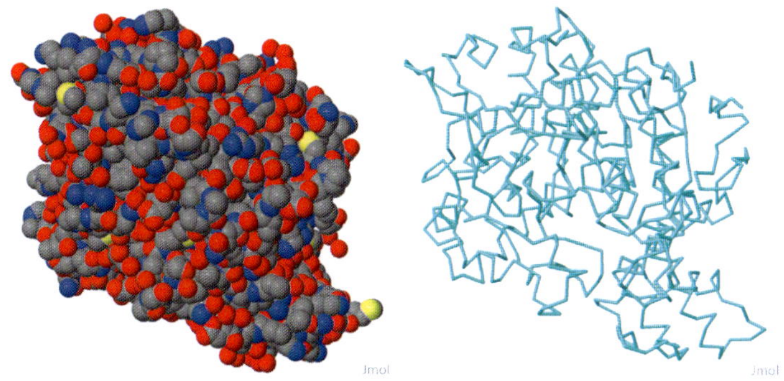


**Figure 6.76** Sphere (left panel a) and backbone (right panel b) representations of the active AChE 1VOT.

(activated) state with different molecules (mainly Tacrine, Rivastigmine, Galantamine, Huperzine A/B).

Figures 6.75 and 6.76 report the sphere (left panel a) and backbone (right panel b) representations of AChE in its native (2ACE) and activated (1VOT) states, respectively, as obtained by the homology modeling procedure.

## 6.9.2 *Topological Properties*

The first task addressed here is to assess the level of resolution the present topological network model can reach when discriminating

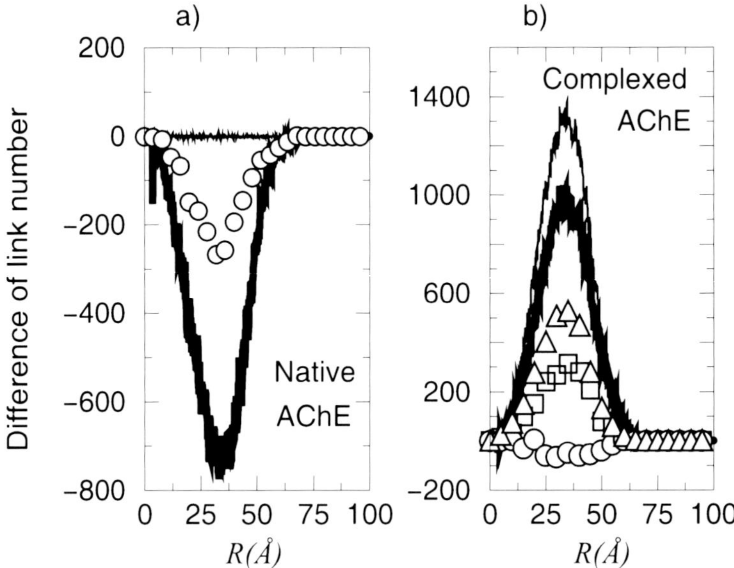


**Figure 6.77** LND for different configurations of AChE versus the interaction radius. All the configurations contain the same number of amino acids. Figure (a): Native configurations. Tiny continuous line refers to the LND between two chains of the 1EA5 representation: 1EA5*b* and 1EA5*a*. Bold continuous line refers to the difference between 1W75*a* and 2ACE. Open circles refer to both: the LND between 1EA5*a* and 1W75*b* and the LND between 1EA5*b* and 1W75*b*. Figure (b): Complexed (active) configurations. Tiny black line refers to the LND between 2ACE and 1ACJ. Bold black line refers to the LND between 2ACE and 1ACL. Open squares refer to the LND between 2ACE and 1AX9. Open circles refer to the LND between 2ACE and 1VOT. Open triangles refer to the LND between 2ACE and 1GPK*a*.

among similar (native or activated) representations and between the native and activated states. In other words, one would check whether the network provides a sensitive map of the protein structure. To this purpose, analysis is carried out by using the AA model.

Figure 6.77 reports the LND between couples of native (left) and complexed (right) representations of AChE as a function of the interaction radius. Here the model correctly predicts that two distinct chains of the same representation of native AChE ($1EA5_{a/b}$) keep practically the same number of links independent of the interaction radius. On the other hand, representations of native AChE obtained under different experimental conditions show significant differences in the number of links, with a maximum value around $R_c = 2$ Å. Some structures of complexed AChE are compared with the native form 2ACE. The difference in the link number can be positive or negative with respect to the considered structure, and the maxima differences are comparable with those between couples of native structures.

If the network model is used to discriminate between the native and activated (or complexed) states of the protein, it is mandatory that the reference representations be produced under the same experimental conditions. For this reason, in the following the couple 2ACE-1VOT-2 (X-ray products, same experiment) is used for native and complexed (with Huperzine A) AChE. The 1VOT-2 structure is the amino acid sequence 1VOT deprived of two amino acids, ALA536 and CYS537, which are not present in 2ACE.

To emphasize the network model's ability to catch the protein topology, the adjacency matrix [Albert and Barabási (2002)] is calculated by representing the links in an $x$–$y$ plane where the serial number of protein amino acids is reported on the $x$- and $y$-axes. Here each link corresponds to a point. Accordingly, Fig. 6.78 reports the contact map of AChE. The adjacency matrix was calculated for $R_c = 6$ Å (dark points), and for $R_c = 12$ Å (grey boxes). Here one should notice an inhomogeneous distribution of links, mainly due to the high complexity of the AChE protein. Within a single figure, it is impossible to report all the connections among sheets and helices. Thus, only some of them should be emphasized, the most evident, mainly reproducing the closeness between $\beta$-sheets.

In conclusion, the graph analysis provides a valuable sketch of the force connected regions in the protein. In fact, by considering short-range forces and using $R_c$ as the parameter describing their cut-off distance, Fig. 6.78 identifies the interacting regions of the protein. The increase in $R_c$ is equivalent to consider forces with longer range. Accordingly, the drawings in Fig. 6.77 emphasize the dependence from the interacting radius of the connective map of the network.

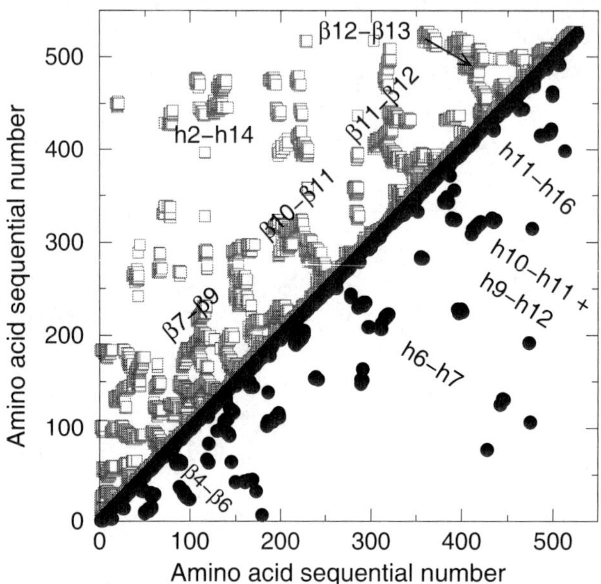


**Figure 6.78** Contact map of 2ACE. The $x, y$ axes report the sequential number of the amino acid. Full circles refer to $R_c = 6$ Å, and open squares refer to $R_c = 12$ Å. Each circle/square corresponds to a link between the couple of amino acids $(x, y)$. Dashed lines are in correspondence with the three amino acids of the activated site.

### 6.9.3 *Small-Signal Electrical Properties*

By means of the Nyquist plot, the electrical responses of the protein in its native and activated states are compared for different values of the interaction radius.

The AA model contains one free parameter in the value of the cut-off radius $R_c$, which fixes the number of links and so the network topology. In the limit of $R_c$ values too small (say 6 Å), only the nearest neighbors are connected, and so it is not possible to reveal the existence of more complex structures such as $\alpha$-helices or $\beta$-sheets. On the other hand, in the limit of $R_c$ values too large (say larger than 80 Å), each node is connected with all the others, and so the protein appears as a uniform structure. A value of $R_c$ that is relevant to discriminate between native and activated states should be that which enables the main structures of the protein to emerge clearly.

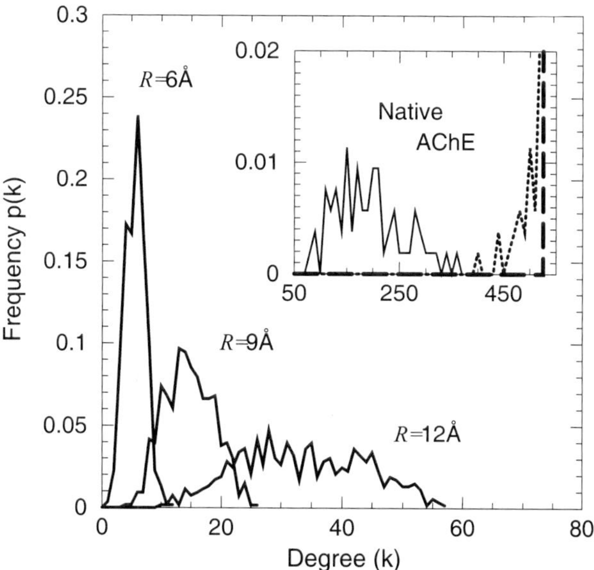


**Figure 6.79** Degree distribution of the native AChE network for increasing values of the interaction radius $R_c$. The inset reports the degree distribution for $R_c = 25$ Å (solid line), $R_c = 50$ Å (dotted line), and $R_c = 80$ Å (bold dashed line).

Indeed, one is interested in detecting if and how they displace in the protein conformational change. Accordingly, one looks for a value of $R_c$ best revealing the main structures of the protein but also emphasizing the differences between the activated and native states of the protein.

To explore the different topologies associated with the changing of $R_c$, in the frame of the AA model, the network degree distribution is evaluated, that is, the distribution of the connected nodes [Albert and Barabási (2002)]. The results of calculations are reported in Fig. 6.79. Here one observes that for $R_c \leq 9$ Å, the degree distribution remains substantially peaked around the same degree value, that is, there is a single characteristic dimension of the network clustering. It corresponds to the nearest neighboring domain ($k = 6$). The cluster dimension grows for increasing $R_c$, until the value $R_c = 9$ Å. One should notice that for the same value of $R_c$, AChE exhibits a degree distribution randomly spiked in the range of $20 <$

$k < 45$. For values of $R_c$ in the range of 12–25 Å, a spreading of the distribution was found, which exhibits a series of spikes representing the fingerprint of the tertiary structure of the given protein. For values of $R_c > 25$ Å, the degree distribution was found to shrink (see insert in Fig. 6.79), and at $R_c = 80$ Å, all the nodes are found to be practically connected to each other. Here the degree distribution takes a delta-like shape centered at $k = (\nu - 1)$. It is concluded that for AChE, $R_c = 9$ Å should be taken as the optimal value to obtain the best contrast of the intimate protein structures.

From the above considerations, four possible cases are discussed in the following when the interaction radius $R_c$ and the effective distance $D$ combine to produce different resolutions for the AA- and AB-directed models, respectively. (Notice that the effective distance for the AB model was assumed to be longer than that for the AA model because of the finite size of the amino acid.)

$$\begin{array}{llll} I. & D_{AA} \approx R,\ D_{AB} \gtrsim R, & D'_{AA} \gtrsim R,\ D'_{AB} \gtrsim R \\ II. & D_{AA} \lesssim R,\ D_{AB} \approx R, & D'_{AA} \gtrsim R,\ D'_{AB} \gtrsim R \\ III. & D_{AA} < R,\ D_{AB} \lesssim R, & D'_{AA} \lesssim R,\ D'_{AB} \gtrsim R \\ IV. & D_{AA} \ll R,\ D_{AB} < R, & D'_{AA} \ll R,\ D'_{AB} < R \end{array} \quad (6.8)$$

By recalling that the condition $D < R$ produces links while $D > R$ does not, the preceding cases are analyzed as follows:

Case I. Here the AA model discriminates different protein states better than the AB model.

Case II. Here both the AA and AB models are able to resolve well the two configurations. In particular, the AA model is more sensitive to the change in the interaction radius.

Case III. Here the AA model discriminates different protein states worse than the AB model.

Case IV. Here it is rather difficult to discriminate the configurations both for the AA and AB models since the number of links remains practically the same in both the configurations.

The general trends discussed above can be quantitatively assessed for the protein under test by selecting a significative set of $R_c$ values 6, 9, 12, and 25 Å. The results of calculations are reported in Appendix with an accuracy of three digits, which is considered appropriate for an experimental validation of the model.

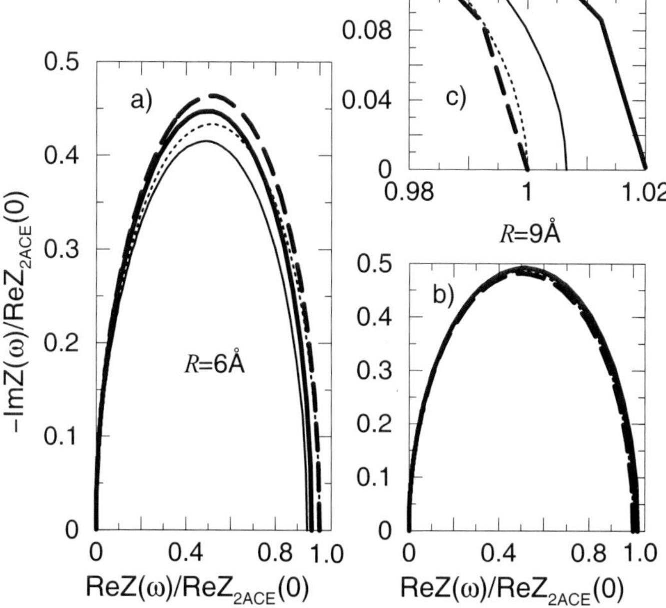


**Figure 6.80** Nyquist plot of the network impedance associated with the AChE protein. Native AChE, 2ACE, is compared with AChE complexed with Huperzine A, 1VOT-2.E. For the AA model, tiny continuous lines refer to 1VOT-2, and dotted lines refer to 2ACE. For the $AB_{\alpha,\beta}$-directed model, bold continuous lines refer to 1VOT-2, and dashed lines refer to 2ACE in the directed $AB_{\alpha,\beta}$ model. In panel (a), $R_c = 6$ Å, and in panels (b) and (c), $R_c = 9$ Å; panel (c) is a zoom of panel (b).

Figures 6.80–6.82 report the Nyquist plots of the global network impedance, normalized to the value at zero frequency, $Z(0)$, for the case of the engineered representations of AChE in the native and activated states. In all the figures, the AA model is compared with the AB models by adopting the same convention for the symbols. In each figure, the plots corresponding to the increasing values of $R_c$ are indicated as (a), (b), and (c) in the clockwise orientation.

As a general trend, the shape of the Nyquist plots remains quite close to that of a semicircle, typical of a single parallel RC impedance, except for small but significant deviations from the semicircle when $R_c = 6$ Å.

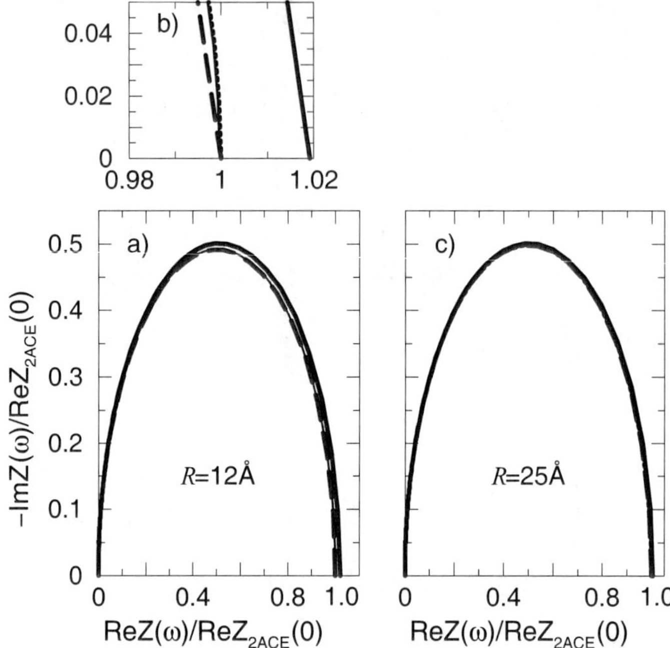


**Figure 6.81** Nyquist plot of the network impedance associated with the AChE protein. Native AChE, 2ACE, is compared with AChE complexed with Huperzine A, 1VOT-2. For the AA model, tiny continuous lines refer to 1VOT-2, and dotted lines refer to 2ACE. For the $AB_{\alpha,\beta}$-directed model, bold continuous lines refer to 1VOT-2, and dashed lines refer to 2ACE in the directed $AB_{\alpha,\beta}$ model. In panels (a) and (b), $R_c = 12$ Å, and panel (b) is a zoom of panel (a). In panel (c), $R_c = 25$ Å.

Figure 6.80 reports the different impedance responses obtained with the AA model (tiny continuous line for the native state and dotted line for the activated state) and with the $AB_{\alpha,\alpha}$-directed model (bold continuous line for the native state and dashed line for the activated one).

Figure 6.81 shows the Nyquist plots for the AA model and the $AB_{\alpha\beta,\alpha\beta}$-directed model, respectively.

Figure 6.82 shows the Nyquist plot for the AA model and the AB isotropic model.

For $R_c = 6$ Å, in all the cases the AA model exhibits a detectable resolution between different configurations. Thereby, the value $R_c$

$= 6$ Å pertains to case I: This value gives a good contrast but not the best.

The value of $R_c = 12$ Åprovides the largest contrast between the activated configuration and the native one.

For $R_c = 25$ Å, the difference between the configurations begins to decrease for both the models, even if the directed AB model still exhibits a resolution increment with respect to that of the AA model.

Finally, for $R_c = 50$ Å, the directed AB model exhibits a resolution increment with respect to that of the AA model.

It should be noticed that the differences between the native and activated configurations are in general small. In the AA model, with $R_c = 6$ Å a difference of only 6% was found. Furthermore, this difference decreases at increasing $R_c$. This implies a low level of resolution between the configurations, even for the most sensitive AB-directed model. Accordingly, Figs. 6.80 and 6.81 report only the comparison between the AA model and the directed $AB_{\alpha,\beta}$ model, which exhibits the best resolution among the AB models.

Figure 6.81 reports the Nyquist plot of AChE for $R_c = 12$ Å and 25 Å, respectively. One should notice that in both cases, the AA model is no longer able to resolve the native from the activated configuration, while the directed AB model resolves a difference between the configurations of 2% for $R_c = 12$ Å, as expected in case III. For $R_c = 25$ Å, also the directed AB model is no longer able to resolve the difference between the configurations, as expected in case IV.

Figure 6.82 reports the fitting of the Nyquist plot for 2ACE, with $R_c = 6$ Å in the AA model, obtained with the Cole–Cole function [Cole and Cole (1941)] and $\alpha$=0.09 together with the ideal semicircle shape corresponding to $\alpha = 0$. The fact that for $R_c > 6$ Å the Nyquist plots take the ideal semicircle shape is explained by the predominant increase of parallel with respect to serial connections. Thus, the network is no longer able to resolve the single relaxation times pertaining to each RC link but exhibits an average time constant.

### 6.9.4 *Conclusion*

By using the topological features of the network, some relevant PDB entries for AChE have been compared in terms of the number of links

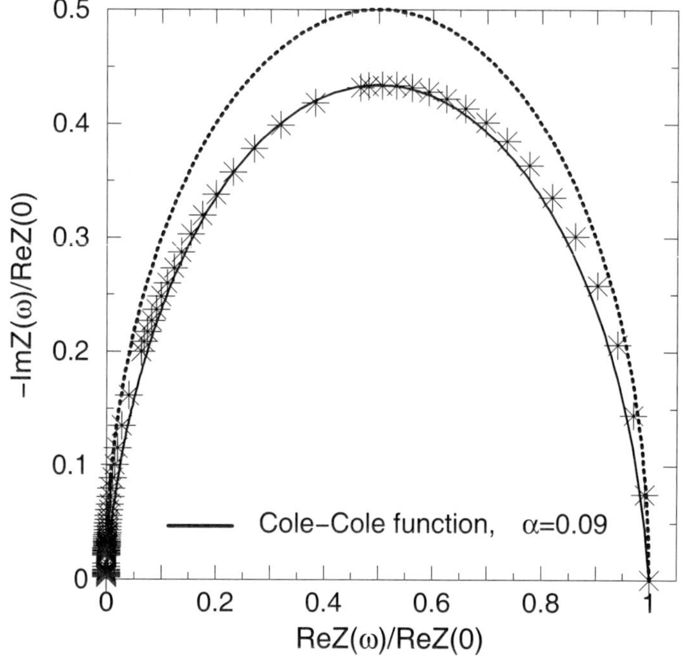


**Figure 6.82** Nyquist plot of 2ACE for $R_c$ = 6 Å, with the Cole–Cole distribution. Stars refer to data from simulations, the dotted line refers to the Debye distribution function, and the continuous line refers to the Cole–Cole distribution function with the best fit parameter $\alpha = 0.09$. The value $\tau = 1.12$ s is used for all the distributions.

as a function of the interacting radius. One should notice that the network map of the protein is able to distinguish among different PDB entries, and also to reproduce, with a fine choice of the $R_c$ value, some topological properties of the protein structures.

From the features of the impedance network associated with the topological one, the dynamic electrical response of the proteins was investigated through the Nyquist plot representation. Accordingly, it was found that the electrical responses of torpedo AChE are quite different in what concerns the possibility to distinguish between the native and activated states. For the couple 2ACE-1VOT-2, the maximal difference is of 6%, and it is obtained for a small value of the interaction radius $R_c$ = 6 Å. Thus, it is concluded that such a small sensitivity is due to an effectively small difference between the two states. Furthermore, since the maximal resolution between the native and activated states is for small values of $R_c$, the conformational change acts only among nearest neighbors. However, this is not the only possibility. As a matter of fact, in a recent work on a particular enzyme, cyclophilin A (CypA) [Eisenmesser et al. (2005)], it was shown that also in the native form, the enzyme lives part in the activated state and part in the native state. In the activated state, the percentage of activated configurations simply increases. According to this observation, also for AChE, present results should demonstrate that the crystallographic image of native AChE is a mix of the native state and of the Huperzine A (in this case) state, and therefore its conformational change is of small relevance.

Chapter 7

# Conclusion and Perspectives

The main conclusions of this monograph can be summarized as follows.

1. Proteotronics is introduced as a discipline investigating the coupling between the protein world (proteomics) and electronics. Proteotronics can be considered a branch of *pantatronics*, a new term announced here to stress the historical development and the overwhelming trend of electronics to be applied to a multitude of different fields of scientific/technological interest (spin, atoms, molecules, mechanics, chemistry, etc.) [Eckel et al. (2014); Pepino et al. (2009)]. The main aim of proteotronics is to propose and achieve innovative electronic devices, based on the selective action of specific proteins [Alfinito et al. (2014)].
2. The correlation between the 3D structure of proteins and their sensing properties is confirmed from an experimental and theoretical point of view.
3. The electrical properties of a single protein are found to be similar to that of a medium-gap semiconductor with Ohmic resistivities estimated in the range of $10^5$ $\Omega$ cm at room temperature.

*Proteotronics: Development of Protein-Based Electronics*
Eleonora Alfinito, Jeremy Pousset, and Lino Reggiani

ISBN 978-981-4613-63-7 (Hardcover), 978-981-4613-64-4 (eBook)
www.panstanford.com

4. The current–voltage characteristic is found to exhibit a superlinear behavior at increasing applied voltages, which is associated with tunneling mechanisms of charge transfer. To this purpose, sequential tunneling between neighborhood amino acids is suggested by comparison between theory and experiments. Typical barrier heights are in the range of 50–300 meV and barrier width below 1 nm.
5. The electrical change associated with the conformational change following a sensing action of the protein is found to be in the range from a few percent up to about a factor of two in magnitude. This variation can be positive or negative, and from a theoretical interpretation, it is found consistent with the change in the spatial location of the amino acids of the single protein. In a conformational change, the variation in the average volume of a protein is estimated to be of a few percent.
6. A microscopic modeling based on an impedance network protein analogous (INPA) is found to provide a unified interpretation of experiments available from literature. Once the tertiary structure of a given protein is given, INPA is useful to visualize the amino acids interconnection through the construction of the contact maps and to predict the electrical properties in a wide range of applied bias. INPA is actually a numerical code based on the Monte Carlo technique and is suitable of further implementation to include advances in the knowledge and characterization of proteins.
7. A systematic investigation of most known transmembrane proteins belonging to the family of G protein–coupled receptors (GPCRs) is reported with the aim of providing a first attempt to the development of a new generation of nanobiosensors able to convert the sensing property manifested by a conformational change into a measurable change of electrical response. A proof of concept on the viability of such an attempt was successfully demonstrated for several sensing proteins (olfactory receptors: rat OR I7, human OR 17–40, light receptors: bacteriorhodopsin, proteorhodopsin, etc.).

Despite the large amount of results available from the literature, the field of proteotronics should be considered still in its infancy. In

**Table 7.1** Summary of OR-production methods for OR-based biosensors

| Methods | Advantages | Disadvantages |
|---|---|---|
| Extracts from tissue or cells | Native structures and functions | Hard to purify specific ORs |
| | Native intracellular connections | Strict storage requirements |
| | Suitable for physical absorption | Need to kill animals |
| Cell-based expression | Nature membrane for ORs | Low expression efficiency |
| | Allow the grafting of tags | Relatively expensive |
| | Single type of ORs | Time consuming |
| Cell-free production | High efficiency and purity | High technique-demanding |
| | Controllable reaction conditions | Relatively high cost |
| Chemical synthesis | Stable secondary structure | Hard to maintain domains |
| | Low cost and high purity | Depend on right sequence |

particular, the availability and handling of the raw material, i.e., the proteins, are still very far from a standard comparable with solid-state and organic materials used in modern nanoelectronics. To this purpose, Table 7.1 [Du et al. (2013)] reports the case of olfactory receptors as a typical example of the present level of production of proteins of wide fundamental and applied interest. Concerning the problem of anchoring the protein to a proper solid substrate, several methodologies have been devised. Accordingly, nanolayers of metal–protein–metal structure, analogue of metal-insulator-metal (MIM) structures, have been successfully realized for the case of bR. Functionalized gold substrate coupled with self-assembled multilayer (SAM) techniques and antibody deposition able to graft the correspondent protein have been developed. Production of thin film, including pR as an active material, has been successfully proposed and realized. Use of field-effect transistors coupled with carbon nanotube technology used to anchor a given protein has been realized.

Concerning experimental techniques for electrical characterization, electrochemical impedance spectroscopy, conductive atomic force microscopy, voltage clamping multichannel, patch clamping single channel, etc. have emerged as powerful and reliable methods [Glatz and Bailey-Hill (2011)]. Other techniques to characterize transducing purposes have also been developed. For example, a short list used for ORs is reported in Table 7.2 [Glatz and Bailey-Hill (2011)].

**Table 7.2** List of optical transduction technologies alternative to the electrical one utilized for olfactory receptor deorphanization and olfactory biosensing, the various techniques used to produce the measurement, and associated literature

| Techniques | References |
|---|---|
| Surface plasmon resonance (SPR) | [Anker et al. (2008)], [Borisov and Wolfbeis (2008)] |
| Fluorescence (including FRET) | [de Kloe et al. (2010)] |
| Luminescence | [Hoa et al. (2007)],[Homola (2003)], [Milligan (2004a)] |
| Bioluminescence (including BRET) | [Roda et al. (2004)], [Santafé et al. (2010)] |
| Chemiluminescence | [Sun et al. (2004)] |
| Absorbance | [Sai et al. (2010)] |

Perspectives of developments in the near future can be summarized as follows.

1. The resolution of the protein 3D structure, especially for both the cases of native and activated states, remains mostly an unsolved issue.
2. The availability of more purified raw materials keeping the sensing features even outside the cell (the so-called in vitro conditions) and reliable protocols of the anchor or deposition technology remains a mandatory issue.
3. The reproducibility and reliability of measurements able to separate the role played by the environment structure used for depositing the protein from that of the protein itself should be implemented significantly to ensure reproducibility of measurements.
4. Microscopic modeling of the folding process once the sequential amino acid structure of the protein is known is still a challenging problem.
5. The identification of the mechanisms of charge transfer still requires confirming and validating test. To this purpose, the roles of temperature as well as of different environments are subjects deserving further intensive research.
6. The determination of the time scales in receptor activation [Gane (2010)] is a wide unsolved problem.
7. The detailed molecular mechanisms by which ligand binding modulates GPCR activity remain poorly understood, despite their widespread use as drug targets.

# Appendix

# Computational Details

This appendix reports more details of the computational procedure used to investigate charge transport properties and associated fluctuations in a given protein already reported in Chapter 4.

## A.1 Calculation of Small-Signal Impedance Spectrum

In the following, the basic steps to calculate the small-signal response of a given protein are briefly summarized.

**A** The input data are constituted by: (i) the backbone 3D structure of the protein (i.e., the coordinates of the $C_\alpha$ atom of each amino acid) in both cases of native and activated states, (ii) the value of the cut-off radius $R_c$, (iii) the value of each amino acid resistivity, (iv) the frequency range of interest, and (v) the geometry of the contacts properly chosen together with the value of the applied bias. The 3D structures are taken from the Protein Data Bank (PDB) or, in its absence, are deduced by the X-ray/NMR/MODELLER files.

**B** The impedance network associated with the protein backbone is obtained by correlating all the couples of amino acids with a distance less than the chosen cut-off value $R_c$.

**C** The elemental impedance associated with each link coupling two neighboring amino acids is calculated (see Eq. 4.4).

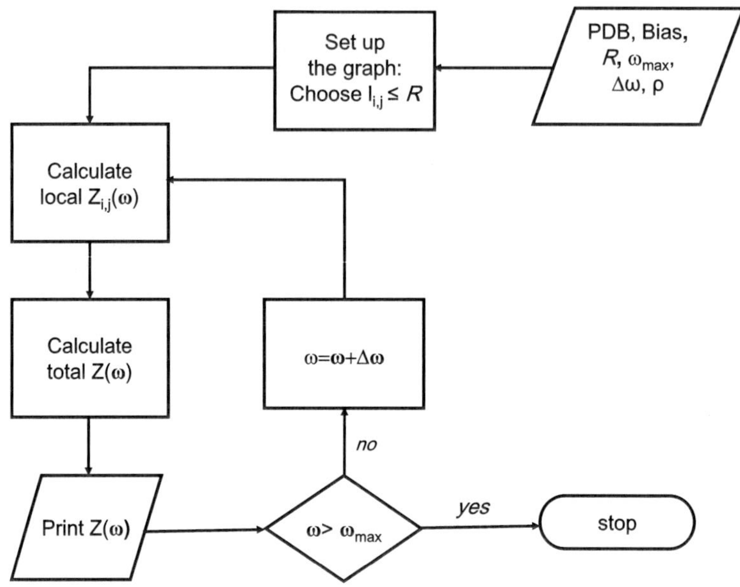


**Figure A.1** Block diagram for the numerical calculation of the small-signal impedance of a single protein.

**D** The total network impedance is calculated by solving the electrical network within the standard Gauss–Jordan method [Press et al. (2006)].

**E** The process is repeated for different frequency values.

The block diagram of this procedure is reported in Fig. A.1.

## A.1.1 *Analysis of the Protein Equivalent Circuit Obtained from Calculations of Bovinerhodopsin and AChE*

The frequency response of the impedance network presented here can be modeled to a good degree of approximation, by the impedance of a single RC parallel circuit:

$$Z_{RC} = \frac{R}{1 + i\omega RC} \equiv \mathcal{R} + \mathcal{I} \tag{A.1}$$

where $\mathcal{R}$, $\mathcal{I}$ indicate the real and imaginary parts of $Z_{RC}$, respectively.

To better compare results coming from different proteins and boundary conditions, it is convenient to normalize the impedance by introducing:

$$\hat{Z}_{RC} = \frac{Z_{RC}}{R} \equiv \hat{\mathcal{R}} + \hat{\mathcal{I}} \tag{A.2}$$

with $R$ the zero-frequency impedance. Notice that Eq. (A.2) defines a first-order function, that is, it has only one pole in frequency and it is stable. The maximum of $-\hat{\mathcal{I}}$ occurs for $-\hat{\mathcal{I}} = \hat{\mathcal{R}} = 1/2$ and is obtained for $\omega = \omega_M = 1/(RC)$.

In the Nyquist plots calculated for the impedance networks, to account for the nonideal semicircle shape, one should determine the frequency $\omega^*$ at which $-\hat{\mathcal{I}} = \hat{\mathcal{R}}$ and define an effective capacitance, $C^*$, by the relation $\omega^* = 1/RC^*$. The difference between the values of $C$ and $C^*$ is a signature of the deviation of the Nyquist plot from the perfect semicircle shape. The values of $R, C$, and $C^*$ corresponding to the Nyquist plots of bovine rhodopsin (BR) and acetylcholinesterase (AChE) reported, respectively, in Section 6.2 and Section 6.9, are summarized in Table A.1 for the case of BR and in Table A.2 for the case of AChE, respectively.

## A.2 Calculations of Intrinsic Fluctuations of the Single-Protein Impedance Due to the Presence of Defects

Intrinsic fluctuations of the network impedance are obtained by allowing for different mechanisms of stochasticity inside the network according to the following approach. The original static network described in the previous sections (also called perfect network) is replaced by a fluctuating network, where some links can be randomly broken and subsequently recovered. To this purpose, two probabilities, $W_b$ and $W_r$, representing the break and the recovery probability for each link, respectively, are defined. Accordingly, each configuration (state) of the random network is characterized by a fraction of broken links, $p$. The value of $p$ at which the network breaks, $p_c$, is a characteristic value ( percolation threshold) related to the network topology. To find the value of $p_c$, the following algorithm is used. First, the perfect network with

**Table A.1** Resistances and capacitances of the RC single-impedance circuit equivalent to the protein impedance network

| | AA Model | | | $AB_{\alpha,\alpha}$ Model | | | $AB_{\alpha\beta,\alpha\beta}$ Model | | |
|---|---|---|---|---|---|---|---|---|---|
| $R_c$ /Type | R ($P\Omega$) | C(fF) | C*(fF) | R($P\Omega$) | C(fF) | C* (fF) | R($P\Omega$) | C(fF) | C*(fF) |
| $R_c$ =6 Å | | | | | | | | | |
| Rho | $9.49 \times 10^3$ | $1.53 \times 10^{-4}$ | $1.32 \times 10^{-4}$ | $4.72 \times 10^3$ | $3.26 \times 10^{-4}$ | $2.82 \times 10^{-4}$ | $4.47 \times 10^3$ | $3.60 \times 10^{-4}$ | $3.19 \times 10^{-4}$ |
| Meta | $1.13 \times 10^4$ | $1.21 \times 10^{-4}$ | $9.31 \times 10^{-5}$ | $5.37 \times 10^3$ | $2.66 \times 10^{-4}$ | $1.86 \times 10^{-4}$ | $5.01 \times 10^3$ | $3.07 \times 10^{-4}$ | $2.50 \times 10^{-4}$ |
| $R_c$ = 12 Å | | | | | | | | | |
| Rho | $2.51 \times 10^2$ | $5.37 \times 10^{-3}$ | $5.16 \times 10^{-3}$ | $1.29 \times 10^2$ | $9.69 \times 10^{-3}$ | $9.69 \times 10^{-3}$ | $9.07 \times 10^1$ | $1.47 \times 10^{-2}$ | $1.43 \times 10^{-2}$ |
| Meta | $3.22 \times 10^2$ | $3.57 \times 10^{-3}$ | $3.41 \times 10^{-3}$ | $1.81 \times 10^2$ | $5.51 \times 10^{-3}$ | $5.01 \times 10^{-3}$ | $1.24 \times 10^2$ | $9.35 \times 10^{-3}$ | $8.93 \times 10^{-3}$ |
| $R_c$=25 Å | | | | | | | | | |
| Rho | $1.67 \times 10^1$ | $6.90 \times 10^{-2}$ | $6.90 \times 10^{-2}$ | $1.07 \times 10^1$ | $1.04 \times 10^{-1}$ | $9.82 \times 10^{-2}$ | 5.99 | $1.99 \times 10^{-1}$ | $1.96 \times 10^{-1}$ |
| Meta | $2.04 \times 10^1$ | $5.38 \times 10^{-2}$ | $5.38 \times 10^{-2}$ | $1.42 \times 10^1$ | $7.04 \times 10^{-2}$ | $7.04 \times 10^{-2}$ | 7.06 | $2.83 \times 10^{-1}$ | $2.83 \times 10^{-1}$ |
| $R_c$=50 Å | | | | | | | | | |
| Rho | 2.31 | $4.32 \times 10^{-1}$ | $4.32 \times 10^{-1}$ | 1.61 | $5.65 \times 10^{-1}$ | $5.65 \times 10^{-1}$ | 0.82 | 1.21 | 1.21 |
| Meta | 2.72 | $3.34 \times 10^{-1}$ | $3.34 \times 10^{-1}$ | 1.97 | $4.24 \times 10^{-1}$ | $4.24 \times 10^{-1}$ | 0.98 | $9.30 \times 10^{-1}$ | $9.30 \times 10^{-1}$ |

*Note*: Model AA is compared with the $AB_{\alpha,\beta}$ and $AB_{\alpha\beta,\alpha\beta}$ models for $R_c = 6, 2, 5$, and 50 Å. Bovine rhodopsin is the analyzed protein.

**Table A.2** Resistances and capacitances of the RC single-impedance circuit equivalent to the protein impedance network

| | AA Model | | | $AB_{\alpha,\beta}$ Model | | |
|---|---|---|---|---|---|---|
| $R_c$/Type | R($P\Omega$) | C(fF) | C*(fF) | R($P\Omega$) | C(fF) | C* (fF) |
| $R_c = 6$ Å | | | | | | |
| 2ACE | $1.51 \times 10^4$ | $7.23 \times 10^{-5}$ | $6.00 \times 10^{-5}$ | $3.63 \times 10^3$ | $3.93 \times 10^{-4}$ | $3.67 \times 10^{-4}$ |
| 1VOT-2 | $1.43 \times 10^4$ | $8.74 \times 10^{-5}$ | $6.99 \times 10^{-5}$ | $3.50 \times 10^3$ | $4.08 \times 10^{-4}$ | $3.57 \times 10^{-4}$ |
| $R_c = 9$ Å | | | | | | |
| 2ACE | $1.45 \times 10^3$ | $1.11 \times 10^{-3}$ | $1.01 \times 10^{-3}$ | $5.76 \times 10^2$ | $2.89 \times 10^{-3}$ | $2.67 \times 10^{-3}$ |
| 1VOT-2 | $1.46 \times 10^3$ | $1.10 \times 10^{-3}$ | $1.07 \times 10^{-3}$ | $5.88 \times 10^2$ | $2.84 \times 10^{-3}$ | $2.62 \times 10^{-3}$ |
| $R_c = 12$ Å | | | | | | |
| 2ACE | $3.57 \times 10^2$ | $5.10 \times 10^{-3}$ | $5.10 \times 10^{-3}$ | $1.42 \times 10^2$ | $1.28 \times 10^{-2}$ | $1.28 \times 10^{-2}$ |
| 1VOT-2 | $3.57 \times 10^2$ | $5.10 \times 10^{-3}$ | $5.10 \times 10^{-3}$ | $1.45 \times 10^2$ | $2.51 \times 10^{-2}$ | $2.51 \times 10^{-2}$ |
| $R_c = 25$ Å | | | | | | |
| 2ACE | $2.08 \times 10^1$ | $9.80 \times 10^{-2}$ | $9.80 \times 10^{-2}$ | 9.71 | $2.07 \times 10^{-1}$ | $2.06 \times 10^{-1}$ |
| 1VOT-2 | $2.08 \times 10^1$ | $9.63 \times 10^{-2}$ | $9.63 \times 10^{-2}$ | 9.76 | $2.05 \times 10^{-1}$ | $2.05 \times 10^{-1}$ |

*Note*: Model AA is compared with the $AB_{\alpha,\alpha}$ and $AB_{\alpha\beta,\alpha\beta}$ models, for $R_c$ = 6, 9, 2, and 25 Å. AChE is the analyzed protein.

all the links active is electrically solved. Second, each link has the possibility to be randomly broken with probability $W_b$, and the updated broken network electrically solved. Third, each broken link has the possibility to be randomly recovered with probability $W_r$, and the updated recovered network electrically solved. Fourth, after the completion of one break-and-recover sweep, the presence of a pathway from one of the contact nodes to the other is checked (to do this, a kind of flood fill algorithm, which is particularly convenient from a computational point of view, is used). If a pathway exists, then the value of $p$ is memorized as $p_i$, where $i$ labels the different realizations, and the process go back to the second step for another realization and so on. If a pathway does not exist, then the value of $p$ is memorized as $p_c$. The average values of $p_i$ and $p_c$ are calculated over many (30,000) realizations. As a validation check of the technique, the same simulations were carried out for a regular square network $20 \times 20$ and they confirmed the result predicted by theory: $< p_c >= 0.5$.

When the average value of the broken links is below the percolation value, the impedance network is stable, but the value of $p$ fluctuates around its average, thus enabling evaluation of the

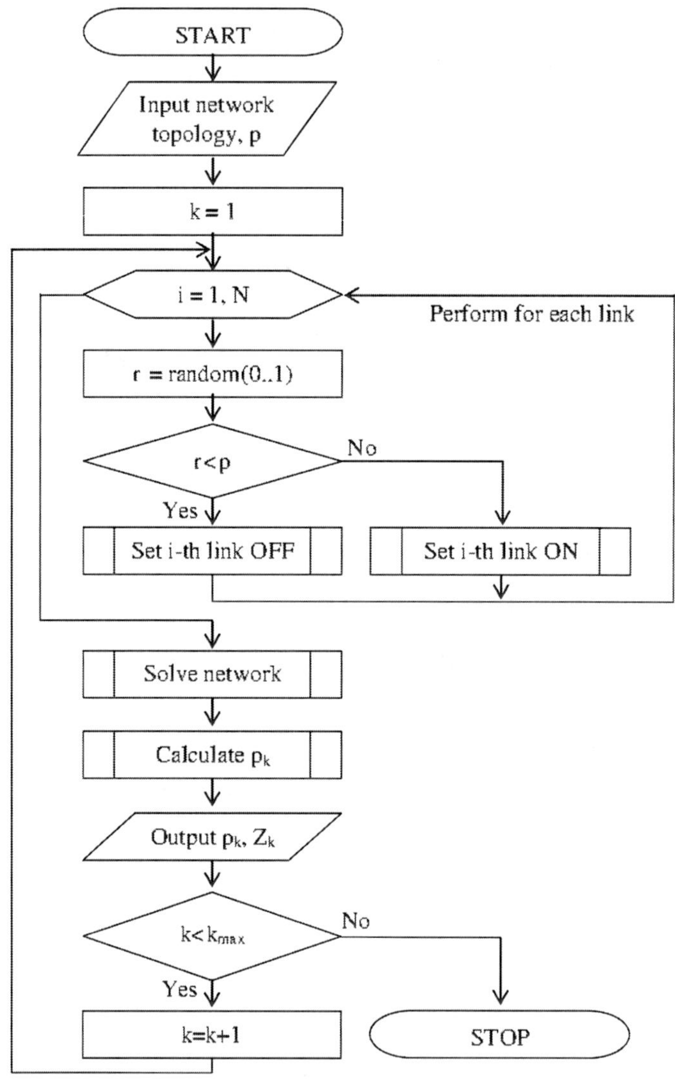


**Figure A.2** Block diagram of a Monte Carlo simulation of the steady state of a random impedance network, including the presence of defects stochastically activated and recovered. The input parameter $p$ is the mean fraction of broken links for the simulated steady state.

intrinsic fluctuations of the impedance associated with the presence of defects. The block diagram of the Monte Carlo simulation of the steady state of a random impedance network, including the presence of defects stochastically activated and recovered, is reported in Fig. A.2.

When a link is broken (open circuit), the nominal value of the corresponding $\rho$ is multiplied by $10^8$ and $\epsilon$ is divided by the same value. In this case, according to Eq. (A.1), the real and imaginary parts of the corresponding elementary impedance are changed by the same quantity.

## A.3 Calculations of Intrinsic Fluctuations of the Single-Protein Impedance Due to Thermal Fluctuations

Two models of thermal fluctuations of the protein structure are considered. The first one, also called the link oscillation model (LOM), is associated with the fluctuations of the distance between nodes around the mean value. The block diagram of the procedure to evaluate the LOM fluctuations is reported in Fig. A.3. The second model, also called the node oscillation model (NOM), is associated with the oscillations of a single node around its average position. The block diagram of the procedure to evaluate the NOM fluctuations is reported in Fig. A.4. A quantum oscillation model, including the possibility of two-force constants, was developed in reference [Pennetta et al. (2005)]. The block diagram of the procedure is reported in Fig. A.5.

## A.4 Calculations of Static High-Field Current–Voltage Characteristics

To account for the super-linear I–V characteristic of a given protein, the model implements a barrier-limited current mechanism as follows [Alfinito et al. (2011a)].

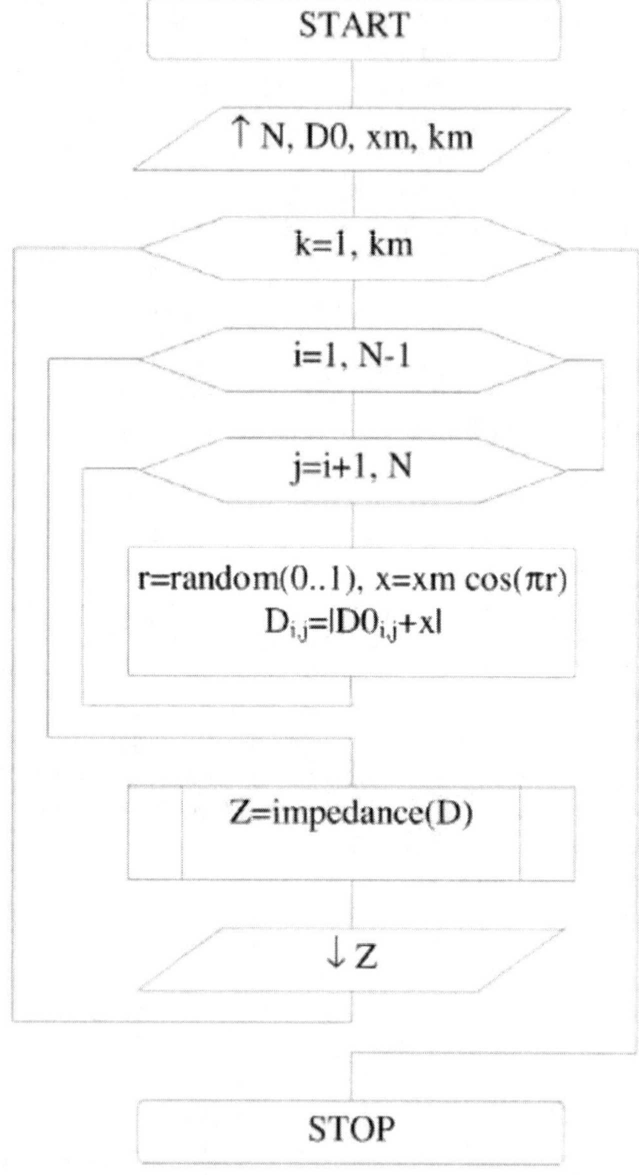


**Figure A.3** Block diagram of LOM. $N$ is the number of nodes, $k_m$ is the number of iterations, $D_0$ is the static matrix of distances, $D$ is the current matrix of distances, and $x_m$ is the amplitude of oscillations.

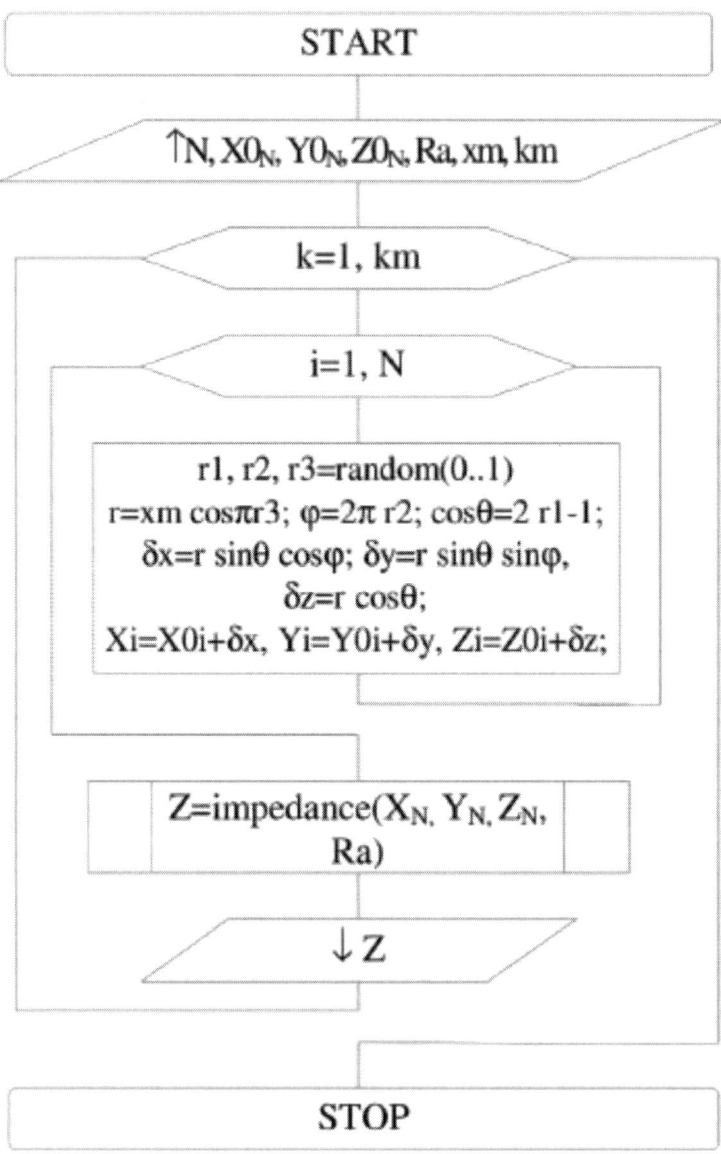


**Figure A.4** Block diagram of NOM. $X_{0N}$, $Y_{0N}$, and $Z_{0N}$ are the static matrices of the coordinates of nodes; $X_N$, $Y_N$, and $Z_N$ are the current matrices of the coordinates of nodes; and $R_a$ is the maximum interaction radius.

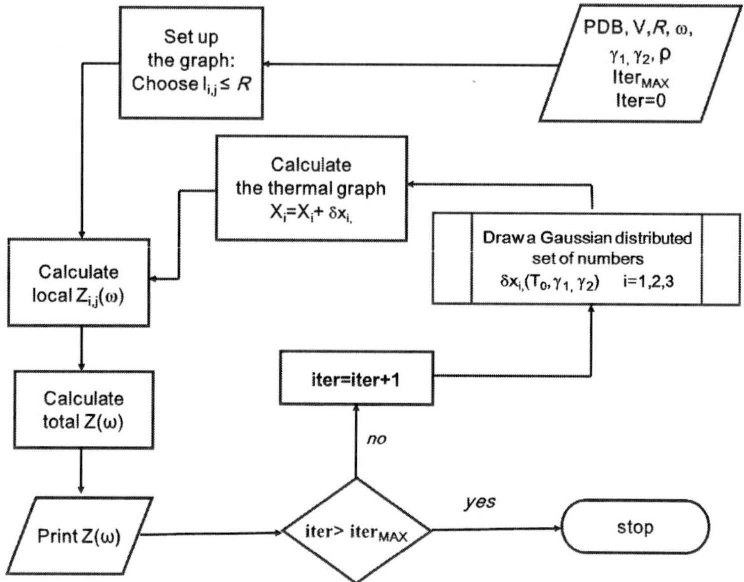


**Figure A.5** Block diagram of the Monte Carlo code for the simulation of the quantum oscillation model.

**A-D** Points from A to D are the same as for the calculation of the small-signal impedance.

**E** At increasing voltages, each elemental resistance is allowed to take a second value of resistivity, $\rho_{min}$, which, playing the role of a small series resistance of the network, is several order of magnitude lower than the original one, $\rho_{max}$. The probability of this choice mimics a barrier-limited mechanism under the direct tunneling (DT) regime, in analogy with the case of an organic molecular layer [Wang et al. (2005)]:

$$P_{i,j} = \exp\left[-\frac{2l_{i,j}}{\hbar}\sqrt{2m(\Phi - eV_{i,j})}\right] \quad \text{(A.3)}$$

where $V_{i,j}$ is the matrix whose elements are the electrical potential drop between the $i$-th and $j$-th nodes calculated with the condition $V_{N,j} = bias$, where N is the total number of amino acids in the protein, and $m$ is an effective electron mass, here taken as that of the free electron, and $\Phi$ is the barrier height. In this way, the initial linear increase in current with applied voltage turns into a super-linear increase, with the value of the barrier energy to be fitted by comparison with experiments.

**F** The value of $\rho$ for each elemental resistance is stochastically determined between the values $\rho_{max}$ and $\rho_{min}$ on the basis of the probability given in Eq. (A.3). In the voltage range where DT regime holds, the deviation from linearity is in general sufficiently weak so that the value chosen for $\rho_{min}$ does not play a significant role. Indeed, the results do not change even after varying this value over two orders of magnitude. On the other hand, the role of the barrier height, which drives the shape of the super-linear behavior, is crucial. The network resistance at the given voltage is then calculated by using the following iterative procedure. First, the network is electrically solved by using the value $\rho_{max}$ for all the elemental resistances. Second, each $\rho_{max}$ is stochastically replaced by $\rho_{min}$ using the probability in Eq. (A.3) according to the local potential drops calculated in the first step; the network is then electrically updated with the new distributed values of $\rho_{max,min}$. Third, the electrical update is iterated (typically $10^4$–$10^5$ iterations depending on the value of the applied voltage) by repeating the second step until the resistance of the network, taken as the average value over the iteration steps, $< R >$, converges within an uncertainty less than 1%. In this third step, the initial iterations (100–2000 depending on voltage) contain a significant numerical noise and as such they are disregarded to avoid an unwanted drift of the average value.

**H** The current at the given voltage is finally calculated as:

$$I = \frac{V}{< R >} \tag{A.4}$$

The block diagram of the above procedure is reported in Fig. A.6.

### A.4.1 *Inclusion of the Fowler–Nordheim Tunneling Mechanism*

At further increasing applied voltages, the super-linear I–V characteristic is associated with a charge transfer between amino acids controlled by a sequential-tunneling mechanism, with the

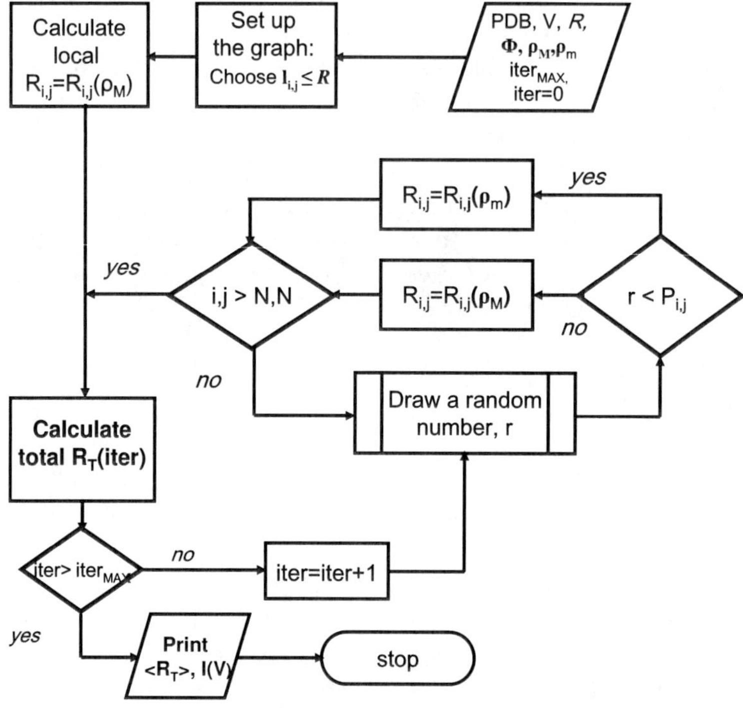


**Figure A.6** Block diagram for the calculation of the super-linear I–V characteristics for a single protein with charge transfer driven by DT regime.

transmission probability being a function of the applied voltage, which interpolates between a direct and an injection (Fowler–Nordheim, FN) tunneling mechanism. In the solution to the resistor network, the tunneling mechanism is implemented by the following procedure. First, the network is electrically solved by using the value $\rho_{max}$ for all the elemental resistances. Second, by using a Monte Carlo acceptance–rejection procedure, each $\rho_{max}$ is stochastically replaced by $\rho_{min}$ using the probability in Eq. (A.6) and Eq. (A.7) according to the local potential drops calculated in the first step. In the high-voltage region ($eV_{i,j} > \Phi$), if the stochastic procedure gives a rejection, then $\rho_{max}$ is replaced by $\rho(V_{i,j})$ of Eq. (A.5). The network is then electrically updated with the new distributed values of $\rho(V_{i,j})$. Third, the electrical update is iterated (typically $10^6$–$10^8$ iterations depending on the value of the applied voltage) by repeating the second step until the value of the network associated current converges within an uncertainty less than a few percent.

To take into account the super-linear response that includes both the DT and the FN tunneling regimes, the link resistivity $\rho$ is chosen to depend on the local voltage drop as:

$$\rho(V) = \begin{cases} \rho_{max}, & eV < \Phi \\ \rho_{max}\left(\frac{\Phi}{eV}\right) + \rho_{min}\left(1 - \frac{\Phi}{eV}\right), & eV \geq \Phi \end{cases} \tag{A.5}$$

Since electron transport is interpreted here in terms of a sequential tunneling between neighboring amino acids, the above interpolation formula reflects the different voltage dependence in the prefactor of the current expression [Wang et al. (2005)]: $I \propto V$ in the DT regime and $I \propto V^2$ in the FN tunneling regime.

For the transmission probability of the tunneling mechanism, the expression given in reference [Casuso et al. (2007a,b); Simmons (1963)] is used:

$$\mathcal{P}^{D}_{i,j} = \exp\left[-\frac{2l_{i,j}}{\hbar}\sqrt{2m\left(\Phi - \frac{1}{2}eV_{i,j}\right)}\right], \qquad eV_{i,j} < \Phi \tag{A.6}$$

$$\mathcal{P}^{FN}_{ij} = \exp\left[-\left(\frac{2l_{i,j}\sqrt{2m}}{\hbar}\right)\frac{\Phi}{eV_{i,j}}\sqrt{\frac{\Phi}{2}}\right], \qquad eV_{i,j} \geq \Phi \tag{A.7}$$

Figure A.7 reports the shape of the tunneling transmission probability that includes both the direct and injection (FN) tunneling regimes (continuous curve) together with that corresponding to DT only (dashed curve).

Figure A.8 reports the voltage dependence of the single-impedance resistivity when tunneling effect is accounted for. The role of the medium (lipids) surrounding the protein is neglected here. Indeed, the medium is not involved in the conformational change, and its effect cancels when considering the variation of the impedance as done here.

The block diagram of the procedure to evaluate the current–voltage characteristics, including a tunneling mechanism with a transmission probability interpolating between direct (low-voltage) and injection (high-voltage) tunneling regimes, is reported in Fig. A.9.

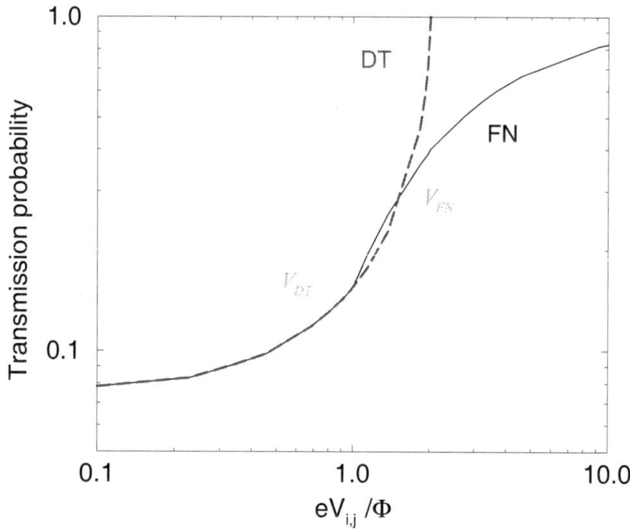


**Figure A.7** Transmission probability as given by DT (dashed line) and the interpolation of DT and FN (continuous line) for the typical parameters: $m_e$ the free electron mass, $l_{i,j} = 5.5$ Å, $\Phi = 219$ meV. The critical voltages, $V_{DT}$ and $V_{FN}$, are explicitly indicated.

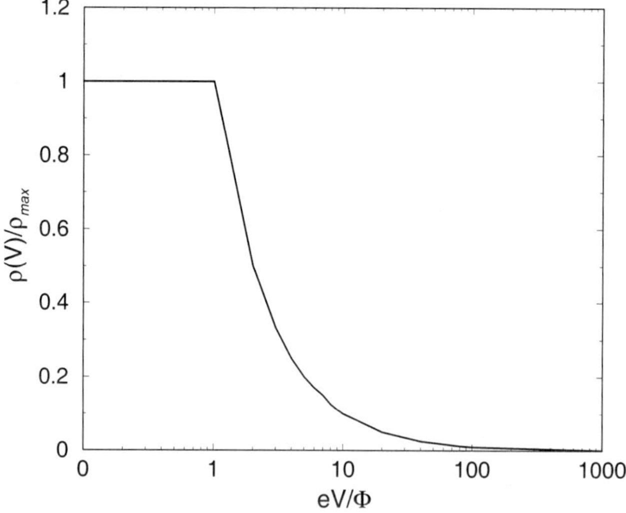


**Figure A.8** Relative resistivity as a function of electrical potential normalized to the barrier energy Φ.

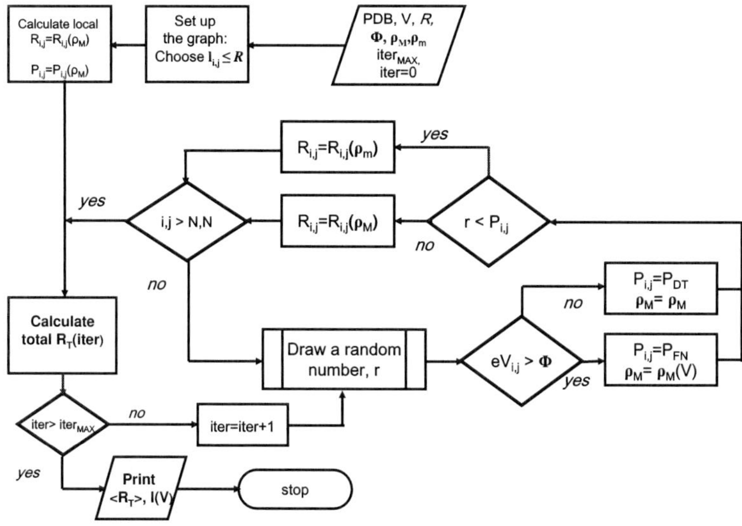


**Figure A.9** Block diagram for the calculation of the I–V characteristics implemented to account for tunneling mechanism, including DT and FN tunneling regimes.

# List of acronyms

AB - Amyl butyrate
AC - Alternate current
AChE - Acetylcholinesterase
AFM - Atomic force microscopy
APTMS - Aminopropyltrimethoxysilane
BLM - Black lipid membrane
BOND - Bioelectronic olfactory neuron device
bR - Bacteriorhodopsin
BR - Bovine rhodopsin
BRET - Bioluminescence resonance energy transfer
c-AFM - Conductive-AFM
CCDS - Consensus coding sequence
CNT - Carbon nanotube
CNTB - Carbon nanotube transistor biosensor
CPE - Constant phase element
CT - Charge transfer
DC - Direct current
DT - Direct tunneling
DUT - Device under test
EC - Extracellular
EIS - Electrochemical impedance spectrometry
EPR - Electron paramagnetic resonance
ET - Exposure time
FET - Field-effect transistor
FICM - Fluorescence interference control microscopy
FLIM - Fluorescence lifetime imaging
FN - Fowler–Nordheim
FRET - Fluorescence resonance energy transfer
GDP - Guanosine diphosphate

GPCR - G protein–coupled receptor
GTP - Guanosine triphosphate
GUV - Giant unilamellar vesicle
HGP - Human genome project
IID - Identical and independently distributed
IN - Impedance network
INPA - Impedance network protein analogous
I–V Current–voltage
LCP - Liquid cubic phase
LND - Link number difference
LOFO - Lift-off float-on
LOM - Link oscillation model
MC - Monte Carlo
MIM - Metal insulator metal
MPM .- Metal protein metal
MSP - Membrane scaffold protein
NMR - Nuclear magnetic resonance
NOM - Node oscillation model
OCB - Olfactory nanovesicle-fused carbon nanotube transistor biosensor
OD - Optical density
OR - Olfactory receptor
PBS - Phosphate buffer solution
PDB - Protein data bank
PDF - Probability distribution function
PM - Purple membrane
pR - Proteorhodopsin
RICM - Reflection interference contrast microscopy
RRV - Relative resistance variation
RSP - Reference sequence project
SAM - Self-assembled multilayer
SEM - Scanning electron microscopy
SLM - Solid lipid membrane
SPOT-NOSED - Single PrOTein Nanobiosensor Grid Array
SPR - Surface plasmon resonance
SUV - Small unilamellar vesicle
SWCNT - Single-walled NCT
TIRFM -Total internal reflection fluorescence microscopy
VFP - Visible fluorescent protein

# Bibliography

Abramowitz, M., and Stegun, I. A. (2012). *Handbook of Mathematical Functions: With Formulas, Graphs, and Mathematical Tables*, (Courier Dover Publications).

Akimov, V., Alfinito, E., Pennetta, C., Reggiani, L., Minic, J., Gorojankina, T., Pajot-Augy, E., and Salesse, R. (2006). An impedance network model for the electrical properties of a single-protein nanodevice, in M. Saraniti and U. Ravaioli (eds.), *Nonequilibrium Carrier Dynamics in Semiconductors, Springer Proceedings in Physics*, Vol. 110 (Springer Berlin Heidelberg), ISBN 978-3-540-36587-7, pp. 229–232, doi: 10.1007/978-3-540-36588-4_51, http://dx.doi.org/10.1007/978-3-540-36588-4_51.

Akkouchi, M. (2005). On the convolution of gamma distributions, *Soochow J. Math.* **31**, 2, pp. 205.

Albert, R., and Barabási, A.-L. (2002). Statistical mechanics of complex networks, *Rev. Mod. Phys.* **74**, pp. 47–97, doi: 10.1103/RevModPhys.74.47, http://link.aps.org/doi/10.1103/RevModPhys.74.47.

Alfinito, E., Akimov, V., Pennetta, C., Reggiani, L., and Gomila, G. (2005). Thermal fluctuations of a GPCR: A two force constant model, *AIP Conf. Proc.* **800**, 1, pp. 381–387, doi: http://dx.doi.org/10.1063/1.2138641, http://scitation.aip.org/content/aip/proceeding/aipcp/10.1063/1.2138641.

Alfinito, E., Millithaler, J.-F., and Reggiani, L. (2012). Gumbel distribution and current fluctuations in critical systems, *Fluct. Noise Lett.* **11**, 03, pp. 1242005.

Alfinito, E., Millithaler, J.-F., Reggiani, L., Zine, N., and Jaffrezic-Renault, N. (2011c). Human olfactory receptor 17-40 as an active part of a nanobiosensor: A microscopic investigation of its electrical properties, *RSC Adv.* **1**, pp. 123–127, doi: 10.1039/C1RA00025J, http://dx.doi.org/10.1039/C1RA00025J.

Alfinito, E., Pennetta, E., and Reggiani, L. (2008). A network model to correlate conformational change and the impedance spectrum of single proteins, *Nanotechnology* **19**, 6, p. 065202, http://stacks.iop.org/0957-4484/19/i=6/a=065202.

Alfinito, E., Millithaler, J. F., Pennetta, C., and Reggiani, L. (2009a). A nanobiosensor based on olfactory receptors, in *Adv. Sens. Interf., 2009. IWASI 2009. 3rd Int. Workshop on*, pp. 25–28, doi: 10.1109/IWASI.2009.5184762.

Alfinito, E., Pennetta, C., and Reggiani, L. (2009b). Smell nanobiosensors: Hybrid systems based on the electrical response to odorant capture theory and experiment, *AIP Conf. Proc.* **1137**, 1, pp. 115–118, doi: http://dx.doi.org/10.1063/1.3156485, http://scitation.aip.org/ content/aip/proceeding/aipcp/10.1063/1.3156485.

Alfinito, E., Pennetta, C., and Reggiani, L. (2009c). Topological change and impedance spectrum of rat olfactory receptor i7: A comparative analysis with bovine rhodopsin and bacteriorhodopsin, *J. Appl. Phys.* **105**, 8, 084703, doi: 10.1063/1.3100210, http://link.aip.org/link/?JAP/105/084703/1.

Alfinito, E., Millithaler, J.-F., and Reggiani, L. (2011a). Charge transport in purple membrane monolayers: A sequential tunneling approach, *Phys. Rev. E* **83**, 4, pp. 042902.

Alfinito, E., Millithaler, J.-F., and Reggiani, L. (2011b). Olfactory receptors for a smell sensor: A comparative study of the electrical responses of rat i7 and human 17-40, *Meas. Sci. Technol.* **22**, 12, p. 124004, http://stacks.iop.org/0957-0233/22/i=12/a=124004.

Alfinito, E., Millithaler, J.-F., Pennetta, C., and Reggiani, L. (2010a). A single protein based nanobiosensor for odorant recognition, *Microelectron. J.* **41**, 11, pp. 718–722, doi: http://dx.doi.org/10.1016/ j.mejo.2010.07.006, http://www.sciencedirect.com/science/article/ pii/S0026269210001278, *IEEE International Workshop on Advances in Sensors and Interfaces 2009.*

Alfinito, E., Pennetta, C., and Reggiani, L. (2010b). Olfactory receptor-based smell nanobiosensors: An overview of theoretical and experimental results, *Sens. Actuators B: Chem.* **146**, 2, pp. 554–558, doi: http:// dx.doi.org/10.1016/j.snb.2010.01.008, http://www.sciencedirect. com/science/article/pii/S0925400510000304, *Selected Papers from the 13th International Symposium on Olfaction and Electronic Nose: ISOEN 2009.*

Alfinito, E., Pousset, J., and Reggiani, L. (2013a). The electrical properties of olfactory receptors in the development of biological smell sensors, in C. J. Crasto (ed.), *Olfactory Receptors, Methods in Molecular Biology*, Vol. 1003 (Humana Press), ISBN 978-1-62703-376-3, pp. 67–83, doi: 10.1007/978-1-62703-377-0_5, http://dx.doi.org/10.1007/978-1-62703-377-0_5.

Alfinito, E., Pousset, J., Reggiani, L., and Lee, K. (2013b). Photoreceptors for a light biotransducer: A comparative study of the electrical responses of two (type-1) opsins, *Nanotechnology* **24**, 39, p. 395501 http://stacks.iop.org/0957-4484/24/i=39/a=395501.

Alfinito, E., and Reggiani, L. (2011a). Charge transport in bacteriorhodopsin monolayers: The contribution of conformational change to current-voltage characteristics, *EPL (Europhys. Lett.)* **85**, 6, p. 68002, http://stacks.iop.org/0295-5075/85/i=6/a=68002.

Alfinito, E., and Reggiani, L. (2009b). Detecting conformational change by current transport in proteins: The case of bacteriorhodopsin monolayers, *J. Phys.: Conf. Series* **193**, 1, p. 012107, http://stacks.iop.org/1742-6596/193/i=1/a=012107.

Alfinito, E., and Reggiani, L. (2013). Role of topology in electrical properties of bacterio-rhodopsin and rat olfactory receptor i7, *Phys. Rev. E* **81**, p. 032902, doi: 10.1103/PhysRevE.81.032902, http://link.aps.org/doi/10.1103/PhysRevE.81.032902.

Alfinito, E., and Reggiani, L. (2013). Evidence of Gumbel distributions ofconductance fluctuations in bacteriorhodopsin thin films, *J. Phys.: Condens. Mat.* **25**, 37, p. 375103, http://stacks.iop.org/0953-8984/25/i=37/a=375103.

Alfinito, E., Reggiani, L., and Pousset, J. (2014). Proteotronics: Electronic devices based on proteins, in *Proceedings of the II National Meeting on Sensors*, http://arxiv.org/abs/1405.3840.

Alfinito, E., and Vitiello, G. (2002). Domain formation in noninstantaneous symmetry-breaking phase transitions, *Phys. Rev. B* **65**, p. 054105, doi: 10.1103/PhysRevB.65.054105, http://link.aps.org/doi/10.1103/PhysRevB.65.054105.

Almen, M. (2009). Mapping the human membrane proteome: A majority of the human membrane proteins can be classified according to function and evolutionary origin, *BMC Biol.* **7**, 1, p. 50.

Andersen, O. S., and Koeppe, R. E. (2007). Bilayer thickness and membrane protein function: An energetic perspective, *Annu. Rev. Biophys. Biomol. Struct.* **36**, 1, pp. 107–130, doi: 10.1146/annurev.biophys.36.040306.132643, http://www.annualreviews.org/doi/abs/10.1146/annurev.biophys.36.040306.132643, pMID: 17263662.

Andolfi, L. (2006). Electron tunneling in a metal–protein–metal junction investigated by scanning tunneling and conductive atomic force spectroscopies, *Appl. Phys. Lett.* **89**, 18, pp. 183125.

Anfinsen, C. B. (1973). Principles that govern the folding of protein chains, *Science* **181**, 4096, pp. 223–230, doi: 10.1126/science.181.4096.223, http://www.sciencemag.org/content/181/4096/223.short.

Anfinsen, C. B., Haber, E., Sela, M., and White, F. H. (1961). The kinetics of formation of native ribonuclease during oxidation of the reduced polypeptide chain, *Proc. Natl. Acad. Sci. USA* **47**, 9, pp. 1309–1314, http://www.pnas.org/content/47/9/1309.short.

Anker, J. N., Hall, W. P., Lyandres, O., Shah, N. C., Zhao, J., and Van Duyne, R. P. (2008). Biosensing with plasmonic nanosensors, *Nat. Mater.* **7**, 6, pp. 442–453, http://dx.doi.org/10.1038/nmat2162.

Antal, T., Labini, F. S., Vasilyev, N. L., and Baryshev, Y. V. (2009). Galaxy distribution and extreme-value statistics, *Europhys. Lett.* **88**, 5, p. 59001, http://stacks.iop.org/0295-5075/88/i=5/a=59001.

Aste, T., and Di Matteo, T. (2008). Emergence of gamma distributions in granular materials and packing models, *Phys. Rev. E* **77**, p. 021309, doi: 10.1103/PhysRevE.77.021309, http://link.aps.org/doi/10.1103/PhysRevE.77.021309.

Aswal, D., Singh, A., Sen, S., Kaur, M., Viswandham, C., Goswami, G., and Gupta, S. (2002). Effect of grain boundaries on paraconductivity of {YBa2Cu3Ox}, *J. Phys. Chem. Solids* **63**, 10, pp. 1797–1803, doi: http://dx.doi. org/10.1016/S0022-3697(01)00266-9, http://www.sciencedirect.com/science/article/pii/S0022369701002669.

Atilgan, A., Durell, S., Jernigan, R., Demirel, M., Keskin, O., and Bahar, I. (2001). Anisotropy of fluctuation dynamics of proteins with an elastic network model, *Biophys. J.* **80**, 1, pp. 505–515, doi: http://dx.doi.org/10.1016/S0006-3495(01)76033-X, http://www.sciencedirect.com/science/article/pii/S000634950176033X.

Bahar, I. (1997). Direct evaluation of thermal fluctuations in proteins using a single-parameter harmonic potential, *Fold. Des.* **2**, 3, pp. 173–181.

Balbín, A., and Andrade, E. (2004). Protein folding and evolution are driven by the maxwell demon activity of proteins, *Acta Biotheoretica* **52**, 3, pp. 173–200, doi: 10.1023/B:ACBI.0000043441.74099.0c, http://dx.doi.org/10.1023/B%3AACBI.0000043441.74099.0c.

Balch, W. E. (1977). An ancient divergence among the bacteria, *J. Mol. Evol.* **9**, 4, pp. 305–311.

Bamann, C., Bamberg, E., Wachtveitl, J., and Glaubitz, C. (2013). Proteorhodopsin, *Biochim. Biophys. Acta-Bioenerg.*, 1837, 5, pp. 614–625, doi: http://dx.doi.org/10.1016/j.bbabio.2013.09.010, http://www.sciencedirect.com/science/article/pii/S0005272813001667.

Barsoukov, E., and Macdonald, J. R. (2005). *Impedance Spectroscopy: Theory, Experiment, and Applications* (Wiley. com).

Barth, A. (2007). Infrared spectroscopy of proteins, *Biochim. Biophys. Acta-Bioenerg.* **1767**, 9, pp. 1073–1101, doi: http://dx.doi.org/10.1016/j.bbabio.2007.06.004, http://www.sciencedirect.com/science/article/pii/S0005272807001375.

Bassler, H. (1993). Charge transport in disordered organic photoconductors a Monte Carlo simulation study, *Phys. Stat. Sol. (b)* **175**, 1, pp. 15–56, http://dx.doi.org/10.1002/pssb.2221750102.

Beja, O., Spudich, E. N., Spudich, J. L., Leclerc, M., and DeLong, E. F. (2001). Proteorhodopsin phototrophy in the ocean, *Nature* **411**, 6839, pp. 786–789, http://dx.doi.org/10.1038/35081051.

Benilova, I., Chegel, V., Ushenin, Y., Vidic, J., Soldatkin, A., Martelet, C., Pajot, E., and Jaffrezic-Renault, N. (2008a). Stimulation of human olfactory receptor 17-40 with odorants probed by surface plasmon resonance, *Eur. Biophys. J.* **37**, 6, pp. 807–814, doi: 10.1007/s00249-008-0272-5, http://dx.doi.org/10.1007/s00249-008-0272-5.

Benilova, I., Vidic, J. M., Pajot-Augy, E., Soldatkin, A., Martelet, C., and Jaffrezic-Renault, N. (2008b). Electrochemical study of human olfactory receptor {OR} 17-40 stimulation by odorants in solution, *Mater. Sci. Eng., C* **28**, 56, pp. 633–639, doi: http://dx.doi.org/10.1016/j.msec.2007.10.040, http://www.sciencedirect.com/science/article/pii/S0928493107001695, *MADICA 2006 Conference, Fifth Maghreb-Europe Meeting on Materials and their Applicatons for Devices and Physical, Chemical and Biological Sensors* MADICA 2006 Conference, Fifth Maghreb-Europe Meeting on Materials and their Applicatons for Devices and Physical, Chemical and Biological Sensors.

Bergo, V., Amsden, J. J., Spudich, E. N., Spudich, J. L., and Rothschild, K. J. (2004). Structural changes in the photoactive site of proteorhodopsin during the primary photoreaction *Biochemistry* **43**, 28, pp. 9075–9083, doi: 10.1021/bi0361968, http://pubs.acs.org/doi/abs/10.1021/bi0361968, pMID: 15248764.

Berman, H. M., Westbrook, J., Feng, Z., Gilliland, G., Bhat, T. N., Weissig, H., Shindyalov, I. N. and Bourne, P. E. (2000). The protein data bank, *Nucleic Acids Res.* **28**, 1, pp. 235–242, doi: 10.1093/nar/28.1.235, http://nar.oxfordjournals.org/content/28/1/235.abstract.

Bertin, E. (2005). Global fluctuations and Gumbel statistics, *Phys. Rev. Lett.* **95**, 17, pp. 170601.

Beyer, M., Menzel, C., Quack, R., Scheper, T., Schgerl, K., Treichel, W., Voigt, H., Ullrich, M., and Ferretti, R. (1994). Development and application of a new enzyme sensor type based on the EIS-capacitance structure for bioprocess control, *Biosens. Bioelectron.* **9**, 1, pp. 17–21, doi: http://dx.doi.org/10.1016/0956-5663(94)80010-3, http://www.sciencedirect.com/science/article/pii/0956566394800103.

Bezanilla, F. (2008). How membrane proteins sense voltage, *Nat. Rev. Mol. Cell Biol.* **9**, 4, pp. 323–332, http://dx.doi.org/10.1038/nrm2376.

Biasco, A. (2004). Self-chemisorption of azurin on functionalized oxide surfaces for the implementation of biomolecular devices, *Mater. Sci. Eng. C, Biomimetic Mater. Sens. Sys.* **24**, 4, pp. 563–567.

Bieri, C., Ernst, O. P., Heyse, S., Hofmann, K. P., and Vogel, H. (1999). Micropatterned immobilization of a G protein-coupled receptor and direct detection of G protein activation, *Nat. Biotech.* **17**, 11, pp. 1105–1108, http://dx.doi.org/10.1038/15090.

Bizzarri, A. R., and Cannistraro, S. (2012). *Dynamic Force Spectroscopy and Biomolecular Recognition* (CRC Press).

BOND (2009–2011). Bioelectronic olfactory neuron device (bond), collaborative project fp7-nmp-2008-small-2, ga number 228685.

Borisov, S. M., and Wolfbeis, O. S. (2008). Optical biosensors, *Chem. Rev.* **108**, 2, pp. 423–461, doi: 10.1021/cr068105t, http://pubs.acs.org/doi/abs/10.1021/cr068105t, pMID: 18229952.

Bourigua, S., Maaref, A., Bessueille, F., and Renault, N. J. (2013). A new design of electrochemical and optical biosensors based on biocatalytic growth of Au nanoparticles example of glucose detection, *Electroanalysis* **25**, 3, pp. 644–651, doi: 10.1002/elan.201200243, http://dx.doi.org/10.1002/elan.201200243.

Bourne, Y. (2003). Structural insights into ligand interactions at the acetylcholinesterase peripheral anionic site, *EMBO J.* **22**, 1, pp. 1–12.

Bramwell, S. T. (2009). The distribution of spatially averaged criticalproperties, *Nat. Phys.* **5**, 6, pp. 444–447.

Branden, T. J., Carl. (1991). *Introduction to Protein Structure* (Garland Publishing, New York, London).

Bucciantini, M., Giannoni, E., Chiti, F., Baroni, F., Formigli, L., Zurdo, J., Taddei, N., Ramponi, G., Dobson, C. M., and Stefani, M. (2002). Inherent toxicity of aggregates implies a common mechanism for protein misfolding diseases, *Nature* **416**, 6880, pp. 507–511, doi: 10.1038/416507a, http://dx.doi.org/10.1038/416507a.

Buck, L., and Axel, R. (1991). A novel multigene family may encode odorant receptors: A molecular basis for odor recognition, *Cell* **65**, 1, pp. 175–187, doi: http://dx.doi.org/10.1016/0092-8674(91)90418-X, http://www.sciencedirect.com/science/article/pii/009286749190418X.

Cai, L., and Zhou, H.-X. (2011). Theory and simulation on the kinetics of protein–ligand binding coupled to conformational change, *J. Chem. Phys.* **134**, 10, 105101, doi: 10.1063/1.3561694, http://link.aip.org/link/?JCP/134/105101/1.

Campi, X. (1986). Multifragmentation: Nuclei break up like percolation clusters, *J. Phys. A: Math. General* **19**, 15, p. L917, http://stacks.iop.org/0305-4470/19/i=15/a=010.

Carloni, P., Rothlisberger, U., and Parrinello, M. (2002). The role and perspective of ab initio molecular dynamics in the study of biological systems, *Acc. Chem. Res.* **35**, 6, pp. 455–464, doi: 10.1021/ar010018u, http://pubs.acs.org/doi/abs/10.1021/ar010018u.

Casuso, I., Fumagalli, L., Samitier, J., Padrs, E., Reggiani, L., Akimov, V., and Gomila, G. (2007a). Electron transport through supported biomembranes at the nanoscale by conductive atomic force microscopy, *Nanotechnology* **18**, 46, p. 465503, http://stacks.iop.org/0957-4484/18/i=46/a=465503.

Casuso, I., Fumagalli, L., Samitier, J., Padrs, E., Reggiani, L., Akimov, V., and Gomila, G. (2007b). Nanoscale electrical conductivity of the purple membrane monolayer, *Phys. Rev. E: Stat. Nonlin. Soft Matter. Phys.* **76**, 4 Pt 1, pp. 041919, http://europepmc.org/abstract/MED/17995038.

Chi, Z., Chen, X. G., Holtz, J. S. W., and Asher, S. A. (1998). UV resonance Raman-selective amide vibrational enhancement: quantitative methodology for determining protein secondary structure, *Biochemistry* **37**, 9, pp. 2854–2864, doi: 10.1021/bi971160z, http://pubs.acs.org/doi/abs/10.1021/bi971160z.

Chiti, F., and Dobson, C. M. (2006). Protein misfolding, functional amyloid, and human disease, *Annu. Rev. Biochem.* **75**, 1, pp. 333–366, doi: 10.1146/annurev.biochem.75.101304.123901, http://www. annualreviews.org/doi/abs/10.1146/annurev.biochem.75.101304.123901, pMID: 16756495.

Choi, A., Kim, S. Y., Kim SY FAU Yoon, S. R., Yoon, S.R., FAU Bae, K., Bae, K., FAU Jung, K.-H., and Kh, J. (2007). Substitution of pro206 and ser86 residues in the retinal binding pocket of anabaena sensory rhodopsin is not sufficient for proton pumping function. *J. Microbiol. Biotechnol.* **17**, 1017-7825 (Linking), pp. 138–145.

Choi, G. (2002). Structural studies of metarhodopsin II, the activated form of the G protein–coupled receptor, rhodopsin, *Biochem. (Easton)* **41**, 23, pp. 7318–7324.

Clusel, M. (2006). Origin of the approximate universality of distributions in equilibrium correlated systems, *Europhys. Lett.* **76**, 6, pp. 1008–1014.

Cole, K. S., and Cole, R. H. (1941). Dispersion and absorption in dielectrics I. Alternating current characteristics, *J. Chem. Phys.* **9**, 4, pp. 341–351, doi: http://dx.doi.org/10.1063/1.1750906, http://scitation.aip.org/content/aip/journal/jcp/9/4/10.1063/1.1750906.

Colovic, M. B., Krstic, D. Z., Lazarevic-Pasti, T. D., Bondzic, A. M., and Vasic, V. M. (2013). Acetylcholinesterase inhibitors: Pharmacology and toxicology, *Curr. Neuropharmacol.* **11**, 3, pp. 315–335, http://www.ingentaconnect.com/content/ben/cn/2013/00000011/00000003/art00006.

Comini, E. (2002). Stable and highly sensitive gas sensors based on semiconducting oxide nanobelts, *Appl. Phys. Lett.* **81**, 10, pp. 1869.

Corcelli, A. (2002). Lipid-protein stoichiometries in a crystalline biological membrane: NMR quantitative analysis of the lipid extract of the purple membrane, *J. Lipid Res.* **43**, 1, pp. 132.

Crasto, C., Singer, M., and Shepherd, G. (2001). The olfactory receptor family album, *Genome Biol.* **2**, 10, pp. 1–4, doi: 10.1186/gb-2001-2-10-reviews1027, http://dx.doi.org/10.1186/gb-2001-2-10-reviews1027.

Cui, Y., Wei, Q., Park, H., and Lieber, C. M. (2001). Nanowire nanosensors for highly sensitive and selective detection of biological and chemical species, *Science* **293**, 5533, pp. 1289–1292, doi: 10.1126/science.1062711, http://www.sciencemag.org/content/293/5533/1289.abstract.

D'Agostino, M. (2003). Critical-like behaviours in central and peripheral collisions: A comparative analysis, *Nucl. Phys. A* **724**, 3-4, pp. 455–476.

Davidson, D. W., and Cole, R. H. (1951). Dielectric relaxation in glycerol, propylene glycol, and npropanol, *J. Chem. Phys.* **19**, 12, pp. 1484–1490, doi: http://dx.doi.org/10.1063/1.1748105, http://scitation.aip.org/content/aip/journal/jcp/19/12/10.1063/1.1748105.

de Kloe, G. E., Retra, K., Geitmann, M., Kallblad, P., Nahar, T., van Elk, R., Smit, A. B., van Muijlwijk-Koezen, J. E., Leurs, R., Irth, H., Danielson, U. H., and de Esch, I. J. P. (2010). Surface plasmon resonance biosensor based fragment screening using acetylcholine binding protein identifies ligand efficiency hot spots (le hot spots) by deconstruction of nicotinic acetylcholine receptor 7 ligands, *J. Med. Chem.* **53**, 19, pp. 7192–7201, doi: 10.1021/jm100834y, http://pubs.acs.org/doi/abs/10.1021/jm100834y.

Debye, P. (1929). Polar molecules. by , ph.d. *New York: Chemical Catalog Co., Inc* **48**, 43, pp. 1036–1037, doi: 10.1002/jctb.5000484320, http://dx.doi.org/10.1002/jctb.5000484320.

Delarue, M. (2002). Simplified normal mode analysis of conformational transitions in DNA-dependent polymerases: The elastic network model, *J. Mol. Biol.* **320**, 5, pp. 1011–1024.

DeLisi, C. (1988). The human genome project: The ambitious proposal to map and decipher the complete sequence of human DNA, *Am. Sci.* **76**, 5, pp. 488–493, http://www.jstor.org/stable/27855388.

DeLong, E. F. (2010). The light-driven proton pump proteorhodopsin enhances bacterial survival during tough times, *PLoS Biol.* **8**, 4, pp. e1000359.

Dioumaev, A. K., Brown, L. S., Shih, J., Spudich, E. N., Spudich, J. L., and Lanyi, J. K. (2002). Proton transfers in the photochemical reaction cycle of proteorhodopsin, *Biochemistry* **41**, 17, pp. 5348–5358, doi: 10.1021/bi025563x, http://pubs.acs.org/doi/abs/10.1021/bi025563x, pMID: 11969395.

Dodd, D., Abraham, D., Eady, D., and Hasnain, S. S. (2000). Structures of oxidized and reduced azurin II from alcaligenes xylosoxidans at 1.75 a resolution, *Acta Crystallogr. Sec. D: Biol. Crystallogr.* **56**, 6, pp. 690–696.

Dorsam, R. T., and Gutkind, J. S. (2007). G protein–coupled receptors and cancer, *Nat. Rev. Cancer* **7**, 2, pp. 79–94, http://dx.doi.org/10.1038/nrc2069.

Du, L., Wu, C., Liu, Q., Huang, L., and Wang, P. (2013). Recent advances in olfactory receptor-based biosensors, *Biosens. Bioelectron.* **42**, pp. 570–580, http://europepmc.org/abstract/MED/23261691.

Duchamp-Viret, P., Chaput, M. A., and Duchamp, A. (1999). Odor response properties of rat olfactory receptor neurons, *Science* **284**, 5423, pp. 2171–2174, doi: 10.1126/science.284.5423.2171, http://www.sciencemag.org/content/284/5423/2171.abstract.

Dufresne, D. (2010). G distributions and the beta-gamma algebra, *Electron. J. Probab.* **15**, pp. 71, 2163–2199, doi: 10.1214/EJP.v15-845, http://ejp.ejpecp.org/article/view/845.

Echenique, P. (2007). Introduction to protein folding for physicists, *Contemporary Phys.* **48**, 2, pp. 81–108, doi: 10.1080/00107510701520843, http://www.tandfonline.com/doi/abs/10.1080/00107510701520843.

Eckel, S., Lee, J. G., Jendrzejewski, F., Murray, N., Clark, C. W., Lobb, C. J., Phillips, W. D., Edwards, M., and Campbell, G. K. (2014). Hysteresis in a quantized superfluid /'atomtronic'/circuit, *Nature* **506**, 7487, pp. 200–203, http://dx.doi.org/10.1038/nature12958.

Eisenmesser, E. Z., Millet, O., Labeikovsky, W., Korzhnev, D. M., Wolf-Watz, M., Bosco, D. A., Skalicky, J. J., Kay, L. E., and Kern, D. (2005). Intrinsic dynamics of an enzyme underlies catalysis, *Nature* **438**, 7064, pp. 117–121, http://dx.doi.org/10.1038/nature04105.

Ferretti, S. (2000). Self-assembled monolayers: A versatile tool for the formulation of bio-surfaces, *Trends Anal. Chem. (Regular ed.)* **19**, 9, pp. 530–540.

Fersht, A. R. (1997). Nucleation mechanisms in protein folding, *Curr. Opin. Struct. Biol.* **7**, 1, pp. 3–9, doi: http://dx.doi.org/10.1016/S0959-440X(97)80002-4, http://www.sciencedirect.com/science/article/pii/S0959440X97800024.

Firestein, S. (2001). How the olfactory system makes sense of scents, *Nature* **413**, 6852, pp. 211–218, http://dx.doi.org/10.1038/35093026.

Fiser, A., and Sali, A. (2003). Modloop: Automated modeling of loops in protein structures, *Bioinformatics* **19**, 18, pp. 2500–2501, doi: 10.1093/bioinformatics/btg362, http://bioinformatics.oxfordjournals.org/content/19/18/2500.abstract.

Fisher, M. E. (1967). The theory of equilibrium critical phenomena, *Rep. Prog. Phys.* **30**, 2, p. 615, http://stacks.iop.org/ 0034-4885/30/i=2/a=306.

Fotiadis, D., Scheuring, S., Mller, S. A., Engel, A., and Mller, D. J. (2002). Imaging and manipulation of biological structures with the {AFM}, *Micron* **33**, 4, pp. 385–397, doi: http://dx.doi.org/10.1016/S0968-4328(01)00026-9, http://www.sciencedirect.com/science/article/pii/S0968432801000269.

Frauenfelder, H., Wolynes, P. G., and Austin, R. H. (1999). Biological Physics, *Rev. Mod. Phys.* **71**, pp. S419–S430, doi: 10.1103/RevModPhys.71.S419, http://link.aps.org/doi/10.1103/RevModPhys.71.S419.

Friedrich, T., Geibel, S., Kalmbach, R., Chizhov, I., Ataka, K., Heberle, J., Engelhard, M., and Bamberg, E. (2002). Proteorhodopsin is a light-driven proton pump with variable vectoriality, *J. Mol. Biol.* **321**, 5, pp. 821–838, doi: http://dx.doi.org/10.1016/S0022-2836(02)00696-4, http://www.sciencedirect.com/science/article/pii/S0022283602006964.

Gaillard, I., Rouquier, S., and Giorgi, D. (2004). Olfactory receptors, *Cell. Mol. Life Sci.* **61**, 4, pp. 456–469, doi: 10.1007/s00018-003-3273-7, http://dx.doi.org/10.1007/s00018-003-3273-7.

Gane, S. (2010). What we do not know about olfaction. Part 1: From nostril to receptor, *Rhinology* **48**, 2, pp. 131–138.

Gardino, A., Villali, J., Kivenson, A., Lei, M., Liu, C., Steindel, P., Eisenmesser, E., Labeikovsky, W., Wolf-Watz, M., Clarkson, M., and Kern, D. (2009). Transient non-native hydrogen bonds promote activation of a signaling protein, *Cell* **139**, 6, pp. 1109–1118, http://europepmc.org/abstract/MED/20005804.

Gardner, J., and Bartlett, P. N.(ed.) (1991). *Sensor and Sensory Systems of an Electronic Nose, Nato Science Series E*, Vol. 212.

Gether, U., and Kobilka, B. K. (1998). G protein-coupled receptors: II. Mechanism of agonist activation, *J. Biol. Chem.* **273**, 29, pp. 17979–17982, doi: 10.1074/jbc.273.29.17979, http://www.jbc.org/content/273/29/17979.short.

Ghanouni, P., Steenhuis, J. J., Farrens, D. L., and Kobilka, B. K. (2001). Agonist-induced conformational changes in the G protein–coupling domain of the ß2 adrenergic receptor, *Proc. Natl. Acad. Sci. USA* **98**, 11, pp. 5997–6002, doi: 10.1073/pnas.101126198, http://www.pnas.org/content/98/11/5997.abstract.

Giraldo, J., and Pin, J.-P. (2011). *G Protein-Coupled Receptors: From Structure to Function,* (RSC Publishing), ISBN 978-1-84973-183-6, doi: 10.1039/9781849733441, http://dx.doi.org/10.1039/9781849733441.

Glatz, R., and Bailey-Hill, K. (2011). Mimicking nature's noses: From receptor deorphaning to olfactory biosensing, *Prog. Neurobiol.* **93**, 2, pp. 270–296, doi: http://dx.doi.org/10.1016/j.pneurobio.2010.11.004, http://www.sciencedirect.com/science/article/pii/S0301008210001954.

Go, N. (1983). Theoretical studies of protein folding, *Annu. Rev. Biophys. Bioeng.* **12**, 1, pp. 183–210, doi: 10.1146/annurev.bb.12.060183.001151, http://www.annualreviews.org/doi/abs/10.1146/annurev.bb.12.060183.001151, pMID: 6347038.

Göpel, W., Ziegler, C., Breer, H., Schild, D., Apfelbach, R., Joerges, J., and Malaka, R. (1998). Bioelectronic noses: A status report part I, *Biosens. Bioelectron.* **13**, 34, pp. 479–493, doi: http://dx.doi.org/10.1016/S0956-5663(97)00092-4, http://www.sciencedirect.com/science/article/pii/S0956566397000924.

Greenfield, N. J., and Fasman, G. D. (1969). Computed circular dichroism spectra for the evaluation of protein conformation, *Biochemistry* **8**, 10, pp. 4108–4116, doi: 10.1021/bi00838a031, http://pubs.acs.org/doi/abs/10.1021/bi00838a031.

Greengard, P. (1976). Possible role for cyclic nucleotides and phosphorylated membrane proteins in postsynaptic actions of neurotransmitters[dagger], *Nature* **260**, 5547, pp. 101–108, http://dx.doi.org/10.1038/260101a0.

Guan, J.-G., Miao, Y.-Q., and Zhang, Q.-J. (2004). Impedimetric biosensors, *J. Biosci. Bioeng.* **97**, 4, pp. 219–226, doi: http://dx.doi.org/10.1016/S1389-1723(04)70195-4, http://www.sciencedirect.com/science/article/pii/S1389172304701954.

Hall, S. E., Floriano, W. B., Vaidehi, N., and Goddard, W. A. (2004). Predicted 3D structures for mouse i7 and rat i7 olfactory receptors and comparison of predicted odor recognition profiles with experiment, *Chem. Sens.* **29**, 7, pp. 595–616, doi: 10.1093/chemse/bjh063, http://chemse.oxfordjournals.org/content/29/7/595.abstract.

Hoa, X., Kirk, A., and Tabrizian, M. (2007). Towards integrated and sensitive surface plasmon resonance biosensors: A review of recent progress, *Biosens. Bioelectron.* **23**, 2, pp. 151–160, doi: http://dx.doi.org/10.1016/j.bios.2007.07.001, http://www.sciencedirect.com/ science/article/pii/S0956566307002710.

Homola, J. (2003). Present and future of surface plasmon resonance biosensors, *Anal. Bioanal. Chem.* **377**, 3, pp. 528–539, doi: 10.1007/s00216-003-2101-0, http://dx.doi.org/10.1007/s00216-003-2101-0.

Horcas, I., Fernandez, R., Gomez-Rodrguez, J. M., Colchero, J., Gomez-Herrero, J., and Baro, A. M. (2007). Wsxm: A software for scanning probe microscopy and a tool for nanotechnology, *Rev. Sci. Instrum.* **78**, 1, 013705, doi: http://dx.doi.org/10.1063/1.2432410, http://scitation.aip.org/content/aip/journal/rsi/78/1/10.1063/1.2432410.

Hou, Y., Helali, S., Zhang, A., Jaffrezic-Renault, N., Martelet, C., Minic, J., Gorojankina, T., Persuy, M.-A., Pajot-Augy, E., Salesse, R., Bessueille, F., Samitier, J., Errachid, A., Akimov, V., Reggiani, L., Pennetta, C., and Alfinito, E. (2006). Immobilization of rhodopsin on a self-assembled multilayer and its specific detection by electrochemical impedance spectroscopy, *Biosens. Bioelectron.* **21**, 7, pp. 1393–1402, doi: http://dx.doi.org/10.1016/j.bios.2005.06.002, http://www.sciencedirect.com/science/article/pii/S095656630500182X.

Hou, Y., Jaffrezic-Renault, N., Martelet, C., Zhang, A., Minic-Vidic, J., Gorojankina, T., Persuy, M.-A., Pajot-Augy, E., Salesse, R., Akimov, V., Reggiani, L., Pennetta, C., Alfinito, E., Ruiz, O., Gomila, G., Samitier, J., and Errachid, A. (2007). A novel detection strategy for odorant molecules based on controlled bioengineering of rat olfactory receptor {I7}, *Biosens. Bioelectron.* **22**, 7, pp. 1550–1555, doi: http://dx.doi.org/10.1016/j.bios.2006.06.018, http://www.sciencedirect.com/science/article/pii/S0956566306002971.

Itzhaki, L. S. (1995). The structure of the transition state for folding of chymotrypsin inhibitor 2 analysed by protein engineering methods: Evidence for a nucleation-condensation mechanism for protein folding, *J. Mol. Biol.* **254**, 2, pp. 260–288.

Jackson, M., and Mantsch, H. (1995). The use and misuse of FTIR spectroscopy in the determination of protein structure, *Crit. Rev. Biochem. Mol. Biol.* **30**, 2, pp. 95–120, doi: 10.3109/10409239509085140, http://informahealthcare.com/doi/abs/10.3109/10409239509085140, pMID: 7656562.

Jacquier, V. (2006). Characterization of an extended receptive ligand repertoire of the human olfactory receptor or 17-40 comprising structurally related compounds functional screening of hor17-40-specific odorants, *J. Neurochem.* **97**, 2, pp. 537–544.

Jacquier, V., Prummer, M., Segura, J.-M., Pick, H., and Vogel, H. (2006). Visualizing odorant receptor trafficking in living cells down to the single-molecule level, *Proc. Natl. Acad. Sci. USA* **103**, 39, pp. 14325–14330, doi: 10.1073/pnas.0603942103, http://www.pnas.org/content/103/39/14325.abstract.

Jaffrezic-Renault, N. (2013). Label-free affinity biosensors based on electrochemical impedance spectroscopy, in S. Marinesco and N. Dale (eds.), *Microelectrode Biosensors, Neuromethods*, Vol. 80 (Humana Press), ISBN 978-1-62703-369-5, pp. 295–318, doi: 10.1007/978-1-62703-370-1_14, http://dx.doi.org/10.1007/978-1-62703-370-1_14.

Jaskierniak, D. (2011). Extracting lidar indices to characterise multilayered forest structure using mixture distribution functions, *Remote Sens. Environ.* **115**, 2, pp. 573–585.

Jianrong, C., Yuqing, M., Nongyue, H., Xiaohua, W., and Sijiao, L. (2004). Nanotechnology and biosensors, *Biotechnol. Adv.* **22**, 7, pp. 505–518, doi: http://dx.doi.org/10.1016/j.biotechadv.2004.03.004, http://www.sciencedirect.com/science/article/pii/S073497500400028X.

Jin, Y., Friedman, N., Sheves, M., and Cahen, D. (2007). Bacteriorhodopsin-monolayer-based planar metal insulator metal junctions via biomimetic vesicle fusion: Preparation, characterization, and bio-optoelectronic characteristics, *Adv. Funct. Mater.* **17**, 8, pp. 1417–1428, doi: 10.1002/adfm.200600545, http://dx.doi.org/10.1002/adfm.200600545.

Jin, Y., Friedman, N., Sheves, M., He, T., and Cahen, D. (2006). Bacteriorhodopsin (bR) as an electronic conduction medium: Current transport through bR-containing monolayers, *Proc. Natl. Acad. Sci. USA* **103**, 23, pp. 8601–8606, http://europepmc.org/abstract/MED/16731629.

Johnson, E. T., Baron, D. B., Naranjo, B., Bond, D. R., Schmidt-Dannert, C., and Gralnick, J. A. (2010). Enhancement of survival and electricity production in an engineered bacterium by light-driven proton pumping, *Appl. Environ. Microbiology* **76**, 13, pp. 4123–4129.

Jones, D., and Reed, R. (1989). Golf: An olfactory neuron specific G protein involved in odorant signal transduction, *Science* **244**, 4906, pp. 790–795, doi: 10.1126/science.2499043, http://www.sciencemag.org/content/244/4906/790.abstract.

Joubaud, S. (2008). Experimental evidence of non-Gaussian fluctuations near a critical point, *Phys. Rev. Lett.* **100**, 18, pp. 180601.

Karplus, M., McCammon, J. A., and Peticolas, W. L. (1981). The internal dynamics of globular protein, *Crit. Rev. Biochem. Mol. Biol.* **9**, 4, pp. 293–349, doi: 10.3109/10409238109105437, http://informahealthcare.com/doi/abs/10.3109/10409238109105437.

Karplus, M., and Weaver, D. (1994). Protein folding dynamics: The diffusion-collision model and experimental data. *Protein Sci.* **3**, 4, pp. 650–668, http://europepmc.org/abstract/MED/8003983.

Katz, E., and Willner, I. (2003). Probing biomolecular interactions at conductive and semiconductive surfaces by impedance spectroscopy: Routes to impedimetric immunosensors, DNA-sensors, and enzyme biosensors, *Electroanalysis* **15**, 11, pp. 913–947, doi: 10.1002/elan.200390114, http://dx.doi.org/10.1002/elan.200390114.

Kendrew, J. C. (1963). Myoglobin and the structure of proteins: Crystallographic analysis and data-processing techniques reveal the molecular architecture, *Science* **139**, 3561, pp. 1259–1266, doi: 10.1126/science. 139.3561.1259, http://www.sciencemag.org/ content/139/3561/1259.short.

Kim, T. H., Lee, S. H., Lee, J., Song, H. S., Oh, E. H., Park, T. H., and Hong, S. (2009). Single-carbon-atomic-resolution detection of odorant molecules using a human olfactory receptor-based bioelectronic nose, *Adv. Mater.* **21**, 1, pp. 91–94, doi: 10.1002/adma.200801435, http://dx.doi.org/10.1002/adma.200801435.

Kitao, A., and Go, N. (1999). Investigating protein dynamics in collective coordinate space, *Curr. Opin. Struct. Biol.* **9**, 2, pp. 164–169, doi: http://dx.doi.org/10.1016/S0959-440X(99)80023-2, http://www.sciencedirect.com/science/article/pii/S0959440X99800232.

Kivioja, J. M., Kurppa, K., Kainlauri, M., Linder, M. B., and Ahopelto, J. (2009). Electrical transport through ordered self-assembled protein monolayer measured by constant force conductive atomic force microscopy, *Appl. Phys. Lett.* **94**, 18, pp. 183901–183903, doi: 10.1063/1.3126448.

Knepp, A. M. (2011). Direct measurement of thermal stability of expressed ccr5 and stabilization by small molecule ligands, *Biochem. (Easton)* **50**, 4, pp. 502–511.

Kobilka, B. K. (2007). G protein coupled receptor structure and activation, *Biochim. Biophys. Acta: Biomembranes* **1768**, 4, pp. 794–807, doi: http://dx.doi.org/10.1016/j.bbamem.2006.10.021, http://www.sciencedirect.com/science/article/pii/S0005273606003981, *G Protein-Coupled Receptors, Signaling Mechanisms and Pathophysiological Relevance*.

Kobilka, B. K., and Deupi, X. (2007). Conformational complexity of G-protein-coupled receptors, *Trends Pharmacol. Sci.* **28**, 8, pp. 397–406, doi: http://dx.doi.org/10.1016/j.tips.2007.06.003, http://www.sciencedirect.com/science/article/pii/S0165614707001459, *Special Issue on Allosterism and Collateral Efficacy*.

Kralj, J. M., Hochbaum, D. R., Douglass, A. D., and Cohen, A. E. (2011). Electrical spiking in Escherichia coli probed with a fluorescent voltage-indicating protein, *Science* **333**, 6040, pp. 345–348, doi: 10.1126/science.1204763, http://www.sciencemag.org/content/333/6040/345.abstract.

Krautwurst, D., Yau, K.-W., and Reed, R. R. (1998). Identification of ligands for olfactory receptors by functional expression of a receptor library, *Cell* **95**, 7, pp. 917–926, doi: http://dx.doi.org/10.1016/S0092-8674(00)81716-X, http://www.sciencedirect.com/science/article/pii/S009286740081716X.

Kubelka, J. (2004). The protein folding speed limit, *Curr. Opin. Struct. Biol.* **14**, 1, pp. 76–88.

Lameh, J., Cone, R., Maeda, S., Philip, M., Corbani, M., Ndasdi, L., Ramachandran, J., Smith, G., and Sade, W. (1990). Structure and function of g protein coupled receptors, *Pharm. Res.* **7**, 12, pp. 1213–1221, doi: 10.1023/A:1015969301407, http://dx.doi.org/10.1023/A%3A1015969301407.

Landau, D., and Binder, K. (2009). *A Guide to Monte Carlo Simulations in Statistical Physics* (Cambridge University Press).

Landau, E. M., and Rosenbusch, J. P. (1996). Lipidic cubic phases: A novel concept for the crystallization of membrane proteins, *Proc. Natl. Acad. Sci. USA* **93**, 25, pp. 14532–14535.

Langmuir, I. (1916). The constitution and fundamental properties of solids and liquids. part I. Solids. *J. Am. Chem. Soc.* **38**, 11, pp. 2221–2295, doi: 10.1021/ja02268a002, http://pubs.acs.org/doi/abs/10.1021/ja02268a002.

Lattanzi, G. (2002). Force dependent transition rates in chemical kinetics models for motor proteins, *J. Chem. Phys.* **117**, 22, pp. 10339.

Launay, G., Sanz, G., Pajot-Augy, E., and Gibrat, J.-F. (2012a). Modeling of mammalian olfactory receptors and docking of odorants, *Biophys. Rev.* **4**, 3, pp. 255–269, doi: 10.1007/s12551-012-0080-0, http://dx.doi.org/10.1007/s12551-012-0080-0.

Launay, G., Téletchéa, S., Wade, F., Pajot-Augy, E., Gibrat, J.-F., and Sanz, G. (2012b). Automatic modeling of mammalian olfactory receptors and docking of odorants, *Protein Eng. Des. Selection* **25**, 8, pp. 377–386, doi: 10.1093/protein/gzs037, http://peds.oxfordjournals.org/ content/25/8/377.abstract.

Lee, B. Y., Sung, M. G., Lee, J., Baik, K. Y., Kwon, Y.-K., Lee, M.-S., and Hong, S. (2011). Universal parameters for carbon nanotube network-based sensors: Can nanotube sensors be reproducible? *ACS Nano* **5**, 6, pp. 4373–4379, doi: 10.1021/nn103056s, http://pubs.acs.org/doi/abs/10.1021/nn103056s.

Lee, M., Im, J., Y., B., Lee, Myung, S., Kang, J., Huang, L., Kwon, Y.-K., and Hong, S. (2006). Linker-free directed assembly of high-performance integrated devices based on nanotubes and nanowires, *Nat. Nano* **1**, 1, pp. 66–71, http://dx.doi.org/10.1038/nnano.2006.46.

Lee, S., and Park, T. (2010). Recent advances in the development of bioelectronic nose, *Biotechnol. Bioprocess Eng.* **15**, 1, pp. 22–29, doi: 10.1007/s12257-009-3077-1, http://dx.doi.org/10.1007/s12257-009-3077-1.

Lee, S. H., Jin, H. J., Song, H. S., Hong, S., and Park, T. H. (2012). Bioelectronic nose with high sensitivity and selectivity using chemically functionalized carbon nanotube combined with human olfactory receptor,

*J. Biotechnol.* **157**, 4, pp. 467–472, doi: http://dx.doi.org/10.1016/j.jbiotec.2011.09.011, http://www.sciencedirect.com/science/article/pii/S0168165611005 487, *Special Issue:* {IBS2010} Part {II} (Biotechnology for a more sustainable environment decontamination and energy production).

Lee, S. H., Jun, S. B., Ko, H. J., Kim, S. J., and Park, T. H. (2009). Cell-based olfactory biosensor using microfabricated planar electrode, *Biosens. Bioelectron.* **24**, 8, pp. 2659–2664, doi: http://dx.doi.org/10.1016/j.bios.2009.01.035, http://www.sciencedirect.com/science/article/pii/S0956566309000530.

Leff, P. (1995). The two-state model of receptor activation, *Trends Pharmacol. Sci.* **16**, 3, pp. 89–97, doi: http://dx.doi.org/ 10.1016/S0165-6147(00)88989-0, http://www.sciencedirect.com/science/article/pii/S0165614700889890.

Lefkowitz, R. J. (2000). The superfamily of heptahelical receptors, *Nat. Cell Biol.* **2**, 7, pp. E133–E136, http://dx.doi.org/10.1038/35017152.

Lefkowitz, R. J. (2004). Historical review: A brief history and personal retrospective of seven-transmembrane receptors, *Trends Pharmacol. Sci. (Regular ed.)* **25**, 8, pp. 413–422.

Leitz, A. J., Bayburt, T. H., Barnakov, A. N., Springer, B. A., and Sligar, S. G. (2006). Functional reconstitution of beta-2-adrenergic receptors utilizing self-assembling nanodisc technology, *Biotechniques* **40**, 5, p. 601.

Levasseur, G., Persuy, M.-A., Grebert, D., Remy, J.-J., Salesse, R., and Pajot-Augy, E. (2003). Ligand-specific dose response of heterologously expressed olfactory receptors, *Eur. J. Biochem.* **270**, 13, pp. 2905–2912, doi: 10.1046/j.1432-1033.2003.03672.x, http://dx.doi.org/10.1046/j.1432-1033.2003.03672.x.

Lidke, D. S., Nagy, P., Heintzmann, R., Arndt-Jovin, D. J., Post, J. N., Grecco, H. E., Jares-Erijman, E. A., and Jovin, T. M. (2004). Quantum dot ligands provide new insights into erbb/her receptor-mediated signal transduction, *Nat. Biotech.* **22**, 2, pp. 198–203, http://dx.doi.org/10.1038/nbt929.

Liu, Q., Cai, H., Xu, Y., Li, Y., Li, R., and Wang, P. (2006). Olfactory cell-based biosensor: A first step towards a neurochip of bioelectronic nose, *Biosens. Bioelectron.* **22**, 2, pp. 318–322, doi: http://dx.doi.org/10.1016/j.bios.2006.01.016, http://www.sciencedirect.com/science/article/pii/S0956566306000315.

Lodish, H., Arnold, B., S Lawrence, Z., Paul, M., David, B., and Darnell., J. (2000). *Mol. Cell Biol., 4th edition* (New York: W. H. Freeman), doi: ISBN-10: 0-7167-3136-3.

Lörinczi, E., Verhoefen, M.-K., Wachtveitl, J., Woerner, A. C., Glaubitz, C., Engelhard, M., Bamberg, E., and Friedrich, T. (2009). Voltage- and ph-dependent changes in vectoriality of photocurrents mediated by wild-type and mutant proteorhodopsins upon expression in xenopus oocytes, *J. Mol. Biol.* **393**, 2, pp. 320–341, doi: http://dx.doi.org/10.1016/j.jmb.2009.07.055, http://www.sciencedirect.com/science/article/pii/S0022283609009073.

Love, J. (2005). Self-assembled monolayers of thiolates on metals as a form of nanotechnology, *Chem. Rev.* **105**, 4, pp. 1103.

Lozier, R., Bogomolni, R., and Stoeckenius, W. (1975). Bacteriorhodopsin: A light-driven proton pump in halobacterium halobium. *Biophys. J.* **15**, 9, pp. 955–962, http://europepmc.org/abstract/MED/1182271.

Luecke, H. (2000). Atomic resolution structures of bacteriorhodopsin photocycle intermediates: The role of discrete water molecules in the function of this light-driven ion pump, *Biochim. Biophys. Acta: Bioenerg.* **1460**, 1, pp. 133–156, doi: http://dx.doi.org/10.1016/S0005-2728(00)00135-3, http://www.sciencedirect.com/science/article/pii/S0005272800001353, *Bacteriorhodopsin*.

Luecke, H., Schobert, B., Richter, H.-T., Cartailler, J.-P., and Lanyi, J. K. (1999). Structure of bacteriorhodopsin at 1.55 resolution, *J. Mol. Biol.* **291**, 4, pp. 899–911, doi: http://dx.doi.org/10.1006/jmbi.1999.3027, http://www.sciencedirect.com/science/article/pii/S0022283699930279.

Macdonald, J. R. (1987). Relaxation in systems with exponential or Gaussian distributions of activation energies, *J. Appl. Phys.* **61**, 2, pp. 700–713, doi: http://dx.doi.org/10.1063/1.338222, http://scitation.aip.org/content/aip/journal/jap/61/2/10.1063/1.338222.

Malnic, B., Hirono, J., Sato, T., and Buck, L. B. (1999). Combinatorial receptor codes for odors, *Cell* **96**, 5, pp. 713–723, doi: http://dx.doi.org/10.1016/S0092-8674(00)80581-4, http://www.sciencedirect.com/science/article/pii/S0092867400805814.

Manickam, A., Johnson, C. A., Kavusi, S., and Hassibi, A. (2012). Interface design for cmos-integrated electrochemical impedance spectroscopy (EIS) biosensors, *Sensors* **12**, 11, pp. 14467–14488, doi: 10.3390/s121114467, http://www.mdpi.com/1424-8220/12/11/14467.

Marrakchi, M., Vidic, J., Jaffrezic-Renault, N., Martelet, C., and Pajot-Augy, E. (2007). A new concept of olfactory biosensor based on interdigitated microelectrodes and immobilized yeasts expressing the human receptor Or 17-40, *Eur. Biophys. J.* **36**, 8, pp. 1015–1018, doi: 10.1007/s00249-007-0187-6, http://dx.doi.org/10.1007/s00249-007-0187-6.

Matteo Pardo, G. S. (ed.) (2009). *Olfaction and Electronic Nose: Proceedings of the 13th International Symposium on Olfaction and Electronic Nose* (American Institute of Physics).

McCusker, E. C., Bane, S. E., O'Malley, M. A., and Robinson, A. S. (2007). Heterologous GPCR expression: A bottleneck to obtaining crystal structures, *Biotechnol. Prog.* **23**, 3, pp. 540–547.

Melikyan, H., Khishigbadrakh, B.-E., Babajanyan, A., Lee, K., Choi, A.-R., Lee, J.-H., Jung, K.-H., and Friedman, B. (2011). Proteorhodopsin characterization based on metal-insulator-metal structure technique, *Thin Solid Films* **519**, 10, pp. 3425–3429, doi: http://dx.doi.org/10.1016/j.tsf.2011.01.250, http://www.sciencedirect.com/science/article/pii/S0040609011003130.

Menon, S. T., Han, M., and Sakmar, T. P. (2001). Rhodopsin: Structural basis of molecular physiology, *Physiol. Rev.* **81**, 4, pp. 1659–1688.

Metzler, R., and Klafter, J. (2002). From stretched exponential to inverse power-law: Fractional dynamics, cole–cole relaxation processes, and beyond, *J. Non-Cryst. Solids* **305**, 13, pp. 81–87, doi: http://dx.doi.org/10.1016/S0022-3093(02)01124-9, http://www.sciencedirect.com/science/article/pii/S0022309302011249.

Milligan, G. (2004a). Applications of bioluminescence- and fluorescence resonance energy transfer to drug discovery at G protein-coupled receptors, *Eur. J. of Pharm. Sci.* **21**, 4, pp. 397–405, doi: http://dx.doi.org/10.1016/j.ejps.2003.11.010, http://www.sciencedirect.com/science/article/pii/S0928098703003300.

Milligan, G. (2004b). G protein-coupled receptor dimerization: Function and ligand pharmacology, *Mol. Pharmacol.* **66**, 1, pp. 1–7, doi: 10.1124/mol.104.000497, http://molpharm.aspetjournals.org/content/66/1/1.abstract.

Minic, J. (2005). Functional expression of olfactory receptors in yeast and development of a bioassay for odorant screening expression of olfactory receptors in yeast for screening, *FEBS J.* **272**, 2, pp. 524–537.

Minic, J., Grosclaude, J., Aioun, J., Persuy, M.-A., Gorojankina, T., Salesse, R., Pajot-Augy, E., Hou, Y., Helali, S., Jaffrezic-Renault, N., Bessueille, F., Errachid, A., Gomila, G., Ruiz, O., and Samitier, J. (2005). Immobilization of native membrane-bound rhodopsin on biosensor surfaces, *Biochim. Biophys. Acta: General Subjects* **1724**, 3, pp. 324–332, doi: http://dx.doi.org/10.1016/j.bbagen.2005.04.017, http://www.sciencedirect.com/science/article/pii/S0304416505001091, *Some Insights into the Biophysics of Complex Systems* International Biophysics Congress.

Miyashita, O., Wolynes, P. G., and Onuchic, J. N. (2005). Simple energy landscape model for the kinetics of functional transitions in proteins, *J. Phys. Chem. B* **109**, 5, pp. 1959–1969, doi: 10.1021/jp046736q, http://pubs.acs.org/doi/abs/10.1021/jp046736q.

Mombaerts, P. (1996). Visualizing an olfactory sensory map, *Cell (Cambridge)* **87**, 4, pp. 675–686.

Mombaerts, P. (2004). Odorant receptor gene choice in olfactory sensory neurons: The one receptor-one neuron hypothesis revisited, *Curr. Opin. Neurobiol.* **14**, 1, pp. 31–36.

Motto, M. G. (1980). Opsin shifts in bovine rhodopsin and bacteriorhodopsin. Comparison of two external point-charge models, *J. Am. Chem. Soc.* **102**, 27, pp. 7947–7949.

Mukhopadhyay, G. S. G., A. Fuchs, and Lay-Ekuakille, A. (eds.) (2010). *A Comparative Study of the Electrical Properties of Rat I7 and Human 17-40 Olfactory Receptors for the Realization of a Nanobiosensor, Proceedings of the IV International Conference on Sensing Technology*.

Mulder, G. J. (1839). Ueber die zusammensetzung einiger thierischen substanzen, *J. Praktische Chemie* **16**, 1, pp. 129–152, doi: 10.1002/prac.18390160137, http://dx.doi.org/10.1002/prac.18390160137.

Nelson, D., and Michael, M. (2004). *Lehninger Principles of Biochemistry, Fourth Edition* (W. H. Freeman).

Neumann, L., Wohland, T., Whelan, R. J., Zare, R. N., and Kobilka, B. K. (2002). Functional immobilization of a ligand-activated G protein–coupled receptor, *ChemBioChem* **3**, 10, pp. 993–998, doi: 10.1002/1439-7633(20021004)3:10<993::AID-CBIC993>3.0.CO;2-Y, http://dx. doi.org/10.1002/1439-7633(20021004)3:10<993::AID-CBIC993>3.0.CO;2-Y.

Nishikawa, T., Murakami, M., and Kouyama, T. (2005) Crystal structure of the 13-cis isomer of bacteriorhodopsin in the dark-adapted state. *J. Mol. Biol.* **352**, pp. 319–328.

Noullez, A. (2002). Global fluctuations in decaying burgers turbulence, *Eur. Phys. J. B, Condens. Mat. Phys.* **28**, 2, pp. 231–241.

Nuzzo, R. G., and Allara, D. L. (1983). Adsorption of bifunctional organic disulfides on gold surfaces, *J. Am. Chem. Soc.* **105**, 13, pp. 4481–4483, doi: 10.1021/ ja00351a063, http://pubs.acs.org/doi/abs/10.1021/ja00351a063.

Ódor, G., and Szolnoki, A. (1996). Directed-percolation conjecture for cellular automata, *Phys. Rev. E* **53**, pp. 2231–2238, doi: 10.1103/PhysRevE.53.2231, http://link.aps.org/doi/10.1103/PhysRevE.53.2231.

Okada, T., Takeda, K., and Kouyama, T. (1998). Highly selective separation of rhodopsin from bovine rod outer segment membranes using combination of divalent cation and alkyl (thio) glucoside, *Photochem. Photobiol.* **67**, 5, pp. 495–499.

Okazaki, K.-I., and Takada, S. (2008). Dynamic energy landscape view of coupled binding and protein conformational change: Induced-fit versus population-shift mechanisms, *Proc. Natl. Acad. Sci.* **105**, 32, pp. 11182–11187, doi: 10.1073/pnas.0802524105, http://www.pnas.org/content/105/32/11182.abstract.

Overhauser, A. W. (1953). Polarization of nuclei in metals, *Phys. Rev.* **92**, pp. 411–415, doi: 10.1103/PhysRev.92.411, http://link.aps.org/doi/10.1103/PhysRev.92.411.

Pace, C. N., and Hermans, J. (1975). The stability of globular protein, *Crit. Rev. Biochem. Mol. Biol.* **3**, 1, pp. 1–43, doi: 10.3109/10409237509102551, http://informahealthcare.com/doi/abs/10.3109/10409237509102551.

Palczewski, K., Kumasaka, T., Hori, T., Behnke, C. A., Motoshima, H., Fox, B. A., Trong, I. L., Teller, D. C., Okada, T., Stenkamp, R. E. et al. (2000). Crystal structure of rhodopsin: Ag protein-coupled receptor, *Sci. Signal.* **289**, 5480, p. 739.

Parak, F. (2003). Proteins in action: The physics of structural fluctuations and conformational changes, *Curr. Opin. Struct. Biol.* **13**, 5, pp. 552–557.

Park, J., Lim, J. H., Jin, H. J., Namgung, S., Lee, S. H., Park, T. H., and Hong, S. (2012). A bioelectronic sensor based on canine olfactory nanovesicle-carbon nanotube hybrid structures for the fast assessment of food quality, *Analyst* **137**, pp. 3249–3254, doi: 10.1039/C2AN16274A, http://dx.doi.org/10.1039/C2AN16274A.

Park, P. S.-H., Lodowski, D. T., and Palczewski, K. (2008). Activation of G protein-coupled receptors: Beyond two-state models and tertiary conformational changes, *Annu. Rev. Pharmacol. Toxicol.* **48**, 1, pp. 107–141, doi: 10.1146/annurev.pharmtox.48.113006.094630, http://www.annualreviews.org/doi/abs/10.1146/annurev.pharmtox.48.113006.094630, pMID: 17848137.

Pebay-Peyroula, E. (1997). X-ray structure of bacteriorhodopsin at 2.5angstroms from microcrystals grown in lipidic cubic phases, *Science (New York)* **277**, 5332, pp. 1676–1681.

Pei, R., Cheng, Z., Wang, E., and Yang, X. (2001). Amplification of antigen-antibody interactions based on biotin labeled protein-streptavidin network complex using impedance spectroscopy, *Biosens. Bioelec-*

*tron.* **16**, 6, pp. 355–361, doi: http://dx.doi.org/10.1016/S0956-5663(01)00150-6, http://www.sciencedirect.com/science/article/pii/S0956566301001506.

Pelton, J. T., and McLean, L. R. (2000). Spectroscopic methods for analysis of protein secondary structure, *Anal. Biochem.* **277**, 2, pp. 167–176, doi: http://dx.doi.org/10.1006/abio.1999.4320, http://www.sciencedirect.com/science/article/pii/S0003269799943208.

Pennetta, C. (2004). Non-Gaussian resistance noise near electrical breakdown in granular materials, *Physica A* **340**, 1-3, pp. 380–387.

Pennetta, C., Akimov, V., Alfinito, E., Reggiani, L., Gomila, G., Ferrari, G., Fumagalli, L., and Sampietro, M. (2005). Modelization of thermal fluctuations in G protein coupled receptors, *AIP Conf. Proc.* **780**, 1, pp. 611–614, doi: http://dx.doi.org/10.1063/1.2036827, http://scitation.aip.org/content/aip/proceeding/aipcp/10.1063/1.2036827.

Pennetta, C., Akimov, V., Alfinito, E., Reggiani, L., Gorojankina, T., Minic, J., Pajot-Augy, E., Persuy, M.-A., Salesse, R., Casuso, I., Errachid, A., Gomila, G., Ruiz, O., Samitier, J., Hou, Y., Jaffrezic, N., Ferrari, G., Fumagalli, L., and Sampietro, M. (2007). *Towards the Realization of Nanobiosensors Based on G protein–coupled Receptors* (Wiley-VCH Verlag GmbH & Co. KGaA), ISBN 9783527610419, doi: 10.1002/9783527610419.ntls0041, http://dx.doi.org/10.1002/9783527610419.ntls0041.

Pennetta, C., Reggiani, L., Trefán, G., and Alfinito, E. (2002). Resistance and resistance fluctuations in random resistor networks under biased percolation, *Phys. Rev. E* **65**, p. 066119, doi: 10.1103/PhysRevE.65.066119, http://link.aps.org/doi/10.1103/PhysRevE.65.066119.

Pennetta, C., Trefán, G., and Reggiani, L. (2000). Scaling law of resistance fluctuations in stationary random resistor networks, *Phys. Rev. Lett.* **85**, pp. 5238–5241, doi: 10.1103/PhysRevLett.85.5238, http://link.aps.org/doi/10.1103/PhysRevLett.85.5238.

Pepino, R. A., Cooper, J., Anderson, D. Z., and Holland, M. J. (2009). Atomtronic circuits of diodes and transistors, *Phys. Rev. Lett.* **103**, p. 140405, doi: 10.1103/PhysRevLett.103.140405, http://link.aps.org/doi/10.1103/PhysRevLett.103.140405.

Pompa, P. P., Biasco, A., Frascerra, V., Calabi, F., Cingolani, R., Rinaldi, R., Verbeet, M. P., de Waal, E., and Canters, G. W. (2004). Solid state protein monolayers: Morphological, conformational, and functional properties, *J. Chem. Phys.* **121**, 21, pp. 10325–10328, doi: http://dx.doi.org/10.1063/1.1828038, http://scitation.aip.org/content/aip/journal/jcp/121/21/10.1063/1.1828038.

Press, W., Flannery, B. P., Teukolsky, S., and Vetterling, W. (2006). Numerical recipes, *Cambridge University Press* **1**, pp. 989–.

Provencher, S. W., and Gloeckner, J. (1981). Estimation of globular protein secondary structure from circular dichroism, *Biochemistry* **20**, 1, pp. 33–37, doi: 10.1021/bi00504a006, http://pubs.acs.org/doi/abs/10.1021/bi00504a006.

Pruitt, K. D., Tatusova, T., Brown, G. R., and Maglott, D. R. (2012). NCBI reference sequences (refseq): Current status, new features and genome annotation policy, *Nucleic Acids Res.* **40**, D1, pp. D130–D135, doi: 10.1093/nar/gkr1079, http://nar.oxfordjournals.org/content/40/D1/D130.abstract.

Quaranta, F., Rella, R., Siciliano, P., Capone, S., Epifani, M., Vasanelli, L., Licciulli, A., and Zocco, A. (1999). A novel gas sensor based on sno2/os thin film for the detection of methane at low temperature, *Sens. Actuators B: Chem.* **58**, 13, pp. 350–355, doi: http://dx.doi.org/10.1016/S0925-4005(99)00095-7, http://www.sciencedirect.com/science/article/pii/S0925400599000957.

Raccosta, S., Baldacchini, C., Rita Bizzarri, A., and Cannistraro, S. (2013). Conductive atomic force microscopy study of single molecule electron transport through the azurin-gold nanoparticle system, *Appl. Phys. Lett.* **102**, 20, 203704, doi: http://dx.doi.org/10.1063/1.4807504, http://scitation.aip.org/content/aip/journal/apl/102/20/10.1063/1.4807504.

Ramachandran, G., Ramakrishnan, C., and Sasisekharan, V. (1963). Stereochemistry of polypeptide chain configurations, *J. Mol. Biol.* **7**, 1, pp. 95–99, doi: http://dx.doi.org/10.1016/S0022-2836(63)80023-6, http://www.sciencedirect.com/science/article/pii/S0022283663800236.

Rebois, R. V., Schuck, P., and Northup, J. K. (2002). Elucidating kinetic and thermodynamic constants for interaction of G protein subunits and receptors by surface plasmon resonance spectroscopy, in J. D. H. Ravi Iyengar (ed.), *G Protein Pathways, Part B: G Proteins and their Regulators, Methods in Enzymology*, Vol. 344 (Academic Press), pp. 15 –42, doi: http://dx.doi.org/10.1016/S0076-6879(02)44703-9, http://www.sciencedirect.com/science/article/pii/S0076687902447039.

Reckel, S., Gottstein, D., Stehle, J., Lhr, F., Verhoefen, M.-K., Takeda, M., Silvers, R., Kainosho, M., Glaubitz, C., Wachtveitl, J., Bernhard, F., Schwalbe, H., Gntert, P., and Dtsch, V. (2011). Solution NMR structure of proteorhodopsin, *Angew. Chem. Int. Ed.* **50**, 50, pp. 11942–11946, doi:

10.1002/anie.201105648, http://dx.doi.org/10.1002/anie.201105648.

Reeves, D. (2008). Membrane mechanics as a probe of ion-channel gating mechanisms, *Phys. Rev. E, Stat. Nonlinear, Soft Mat. Phys.* **78**, 4, p. 041901.

Reuveni, S., Granek, R., and Klafter, J. (2008). Proteins: Coexistence of stability and flexibility, *Phys. Rev. Lett.* **100**, p. 208101, doi: 10.1103/ PhysRevLett.100.208101, http://link.aps.org/doi/10.1103/PhysRevLett.100.208101.

Rhodes, G. (2006). *Crystallography Made Crystal Clear (Third Edition)* (Academic Press).

Rickert, J., Gpel, W., Beck, W., Jung, G., and Heiduschka, P. (1996). A mixed self-assembled monolayer for an impedimetric immunosensor, *Biosens. Bioelectron.* **11**, 8, pp. 757–768, doi: http://dx.doi.org/10.1016/0956-5663(96)85927-6, http://www.sciencedirect.com/science/article/pii/0956566396859276.

Riesenfeld, C. S. (2004). Metagenomics: Genomic analysis of microbial communities, *Annu. Rev. Genet.* **38**, 1, pp. 525–552.

Roda, A., Pasini, P., Mirasoli, M., Michelini, E., and Guardigli, M. (2004). Biotechnological applications of bioluminescence and chemiluminescence, *Trends Biotechnol.* **22**, 6, pp. 295–303, doi: http://dx.doi.org/10.1016/j.tibtech.2004.03.011, http://www.sciencedirect.com/science/article/pii/S016777990400085X.

Rodger, A., and Nordén, B. (1997). *Circular Dichroism and Linear Dichroism* (Book series: Oxford Chemistry Masters).

Ron, I., Sepunaru, L., Itzhakov, S., Belenkova, T., Friedman, N., Pecht, I., Sheves, M., and Cahen, D. (2010). Proteins as electronic materials: Electron transport through solid-state protein monolayer junctions, *J. Am. Chem. Soc.* **132**, 12, pp. 4131–4140, doi: 10.1021/ja907328r, http://pubs.acs.org/doi/abs/10.1021/ja907328r, pMID: 20210314.

Roy, A., Kucukural, A., and Zhang, Y. (2010). I-tasser: A unified platform for automated protein structure and function prediction. *Nat. Protoc.* **5**, 4, pp. 725–738, http://europepmc.org/abstract/MED/20360767.

Sabehi, G. (2004). Different sar86 subgroups harbour divergent proteorhodopsins, *Environ. Microbiol.* **6**, 9, pp. 903–910.

Sai, V., Kundu, T., Deshmukh, C., Titus, S., Kumar, P., and Mukherji, S. (2010). Label-free fiber optic biosensor based on evanescent wave absorbance at 280 nm, *Sens. Actuators B: Chem.* **143**, 2, pp. 724–730.

Sali, A., and Blundell, T. L. (1990). Definition of general topological equivalence in protein structures. A procedure involving comparison of properties and relationships through simulated annealing and dynamic programming. *J. Mol. Biol.* **212**, 2, pp. 403–428, http://view.ncbi.nlm.nih.gov/pubmed/2181150.

Samama, P., Cotecchia, S., Costa, T., and Lefkowitz, R. J. (1993). A mutation-induced activated state of the beta 2-adrenergic receptor. Extending the ternary complex model. *J. Biol. Chem.* **268**, 7, pp. 4625–4636, http://www.jbc.org/content/268/7/4625.abstract.

Samanta, D., and Sarkar, A. (2011). Immobilization of bio-macromolecules on self-assembled monolayers: Methods and sensor applications, *Chem. Soc. Rev.* **40**, pp. 2567–2592, doi: 10.1039/C0CS00056F, http://dx.doi.org/10.1039/C0CS00056F.

Sampietro, M., Ferrari, G., and Natali, D. (2005). Broadband and low noise integrator circuit, Patent WO2005062061AI.

Santafé, A. A.-M., Blum, L. J., Marquette, C. A., and Girard-Egrot, A. P. (2010). Chelating Langmuir–Blodgett film: A new versatile chemiluminescent sensing layer for biosensor applications, *Langmuir* **26**, 3, pp. 2160–2166, doi: 10.1021/la902652d, http://pubs.acs.org/doi/abs/10.1021/la902652d, pMID: 20000740.

Sato, K., Pellegrino, M., Nakagawa, T., Nakagawa, T., Vosshall, L., and Touhara, K. (2008). Insect olfactory receptors are heteromeric ligand-gated ion channels. *Nature* **452**, 7190, pp. 1002–1006, http://europepmc.org/abstract/MED/18408712.

Serebryany, E., Zhu, G. A., and Yan, E. C. (2012). Artificial membrane-like environments for *in vitro* studies of purified G protein–coupled receptors, *Biochim. Biophys. Acta: Biomembranes* **1818**, 2, pp. 225–233.

Sheu, S.-Y. (2002). Charge transport in a polypeptide chain, *Eur. Phys. J. D, At., Mol., Opt. Phys.* **20**, 3, pp. 557–563.

Simmons, J. G. (1963). Generalized formula for the electric tunnel effect between similar electrodes separated by a thin insulating film, *J. Appl. Phys.* **34**, 6, pp. 1793–1803, doi: http://dx.doi.org/10.1063/1.1702682, http://scitation.aip.org/content/aip/journal/jap/34/6/10.1063/1.1702682.

Song, X. (2002). An inhomogeneous model of protein dielectric properties: Intrinsic polarizabilities of amino acids, *J. Chem. Phys.* **116**, 21, pp. 9359–9363, doi: http://dx.doi.org/10.1063/1.1474582, http://scitation.aip.org/content/aip/journal/jcp/116/21/10.1063/1.1474582.

Spinozzi, F., and Beltramini, M. (2012). Quafit: A novel method for the quaternary structure determination from small-angle scattering data, *Biophys. J.* **103**, 3, pp. 511–521, doi: http://dx.doi.org/10.1016/j.bpj.2012.06.037, http://www.sciencedirect.com/science/article/pii/S0006349512007266.

SPOT-NOSED (2003–2005). Single protein nanobiosensor grid array (spot-nosed, project ist-2001-38899 of e. c.).

Standfuss, J. (2011). The structural basis of agonist-induced activation in constitutively active rhodopsin, *Nature (London)* **471**, 7340, pp. 656–660.

Stauffer, D., and Aharony, A. (1991). *Introduction to Percolation Theory; Rev. Version* (Taylor and Francis, London).

Storri, S., Santoni, T., Minunni, M., and Mascini, M. (1998). Surface modifications for the development of piezoimmunosensors, *Biosens. Bioelectron.* **13**, 34, pp. 347–357, doi: http://dx.doi.org/10.1016/S0956-5663(97)00119-X, http://www.sciencedirect.com/science/article/pii/S095656639700119X.

Sun, C., Yang, J., Li, L., Wu, X., Liu, Y., and Liu, S. (2004). Advances in the study of luminescence probes for proteins, *J. Chromatogr. B* **803**, 2, pp. 173–190, doi: http://dx.doi.org/10.1016/j.jchromb.2003.12.039, http://www.sciencedirect.com/science/article/pii/S157002 3203010481.

Sutto, L., Tiana, G., and Broglia, R. A. (2006). Sequence of events in folding mechanism: Beyond the G model, *Protein Sci.* **15**, 7, pp. 1638–1652, doi: 10.1110/ps.052056006, http://dx.doi.org/10.1110/ps.052056006.

Tirion, M. M. (1996). Large amplitude elastic motions in proteins from a single-parameter, atomic analysis, *Phys. Rev. Lett.* **77**, pp. 1905–1908, doi: 10.1103/PhysRevLett.77.1905, http://link.aps.org/doi/10.1103/PhysRevLett.77.1905.

Tokumasu, F., Jin, A. J., and Dvorak, J. A. (2002). Lipid membrane phase behaviour elucidated in real time by controlled environment atomic force microscopy, *J. Electron Microsc.* **51**, 1, pp. 1–9.

Ulman, A. (1996). Formation and structure of self-assembled monolayers, *Chem. Rev.* **96**, 4, pp. 1533–1554, doi: 10.1021/cr9502357, http://pubs.acs.org/doi/abs/10.1021/cr9502357.

Vaidehi, N., Floriano, W. B., Trabanino, R., Hall, S. E., Freddolino, P., Choi, E. J., Zamanakos, G., and Goddard, W. A. (2002). Prediction of structure and function of G protein-coupled receptors, *Proc. Natl. Acad. Sci. USA* **99**, 20, pp. 12622–12627, doi: 10.1073/pnas.122357199, http://www.pnas.org/content/99/20/12622.abstract.

Vericat, C., Vela, M., Benitez, G., Carro, P., and Salvarezza, R. (2010). Self-assembled monolayers of thiols and dithiols on gold: New challenges for a well-known system, *Chem. Soc. Rev.* **39**, 5, pp. 1805–1834.

Vestergaard, M. D., Kerman, K., and Tamiya, E. (2007). An overview of label-free electrochemical protein sensors, *Sensors* **7**, 12, pp. 3442–3458, doi: 10.3390/s7123442, http://www.mdpi.com/1424-8220/7/12/3442.

Vickery, H. B. (1950). The origin of the word protein, *Yale J. Biol. Med.* **22(5)**, p. 387–393.

Vidic, J., Grosclaude, J., Monnerie, R., Persuy, M.-A., Badonnel, K., Baly, C., Caillol, M., Briand, L., Salesse, R., and Pajot-Augy, E. (2008). On a chip demonstration of a functional role for odorant binding protein in the preservation of olfactory receptor activity at high odorant concentration, *Lab Chip* **8**, pp. 678–688, doi: 10.1039/B717724K, http://dx.doi.org/10.1039/B717724K.

Vidic, J. M., Grosclaude, J., Persuy, M.-A., Aioun, J., Salesse, R., and Pajot-Augy, E. (2006). Quantitative assessment of olfactory receptors activity in immobilized nanosomes: A novel concept for bioelectronic nose, *Lab Chip* **6**, pp. 1026–1032, doi: 10.1039/B603189G, http://dx.doi.org/10.1039/B603189G.

Vo-Dinh, T., Cullum, B. M., and Stokes, D. L. (2001). Nanosensors and biochips: Frontiers in biomolecular diagnostics, *Sens. Actuators B: Chem.* **74**, 13, pp. 2–11, doi: http://dx.doi.org/10.1016/S0925-4005(00)00705-X, http://www.sciencedirect.com/science/article/pii/S092540050000705X, *Proceedings of the 5th European Conference on Optical Chemical Sensors and Biosensors.*

Váró, G., and Keszthelyi, L. (1983). Photoelectric signals from dried oriented purple membranes of halobacterium halobium, *Biophys. J.* **43**, 1, pp. 47–51, http://linkinghub.elsevier.com/retrieve/pii/S0006349583843227.

Wallrabe, H. (2005). Imaging protein molecules using FRET and film microscopy, *Curr. Opin. Biotechnol.* **16**, 1, pp. 19–27.

Wallrabe, H., Elangovan, M., Burchard, A., Periasamy, A., and Barroso, M. (2002). Fret microscopy reveals clustered distribution of co-internalized receptor-ligand complexes in the apical recycling endosome of polarized epithelial mdck cells, pp. 64–72, doi: 10.1117/12.470677, http://dx.doi.org/10.1117/12.470677.

Wang, W., Lee, T., and Reed, M. (2005). Electron tunnelling in self-assembled monolayers, *Reports Prog. Phys.* **68**, 3, pp. 523–544.

Weil, J. A., and Bolton, J. R. (2006). *Electron Paramagnetic Resonance: Elementary Theory and Practical Applications* (John Wiley & Sons,

Inc.), ISBN 9780470084984, doi: 10.1002/9780470084984.fmatter, http://dx.doi.org/10.1002/9780470084984.fmatter.

Weiss, T. F. (1997). *Cellular Biophysics* (The MIT Press).

Wesolowska, O., Michalak, K., Maniewska, J., and Hendrich, A. B. (2009). Giant unilamellar vesicles: A perfect tool to visualize phase separation and lipid rafts in model systems, *Acta Biochim. Polonica* **56**, 1.

Wishart, D. S., Sykes, B. D., and Richards, F. M. (1992). The chemical shift index: A fast and simple method for the assignment of protein secondary structure through NMR spectroscopy, *Biochemistry* **31**, 6, pp. 1647–1651, doi: 10.1021/bi00121a010, http://pubs.acs.org/doi/abs/10.1021/bi00121a010.

Woese, C. R., Kandler, O., and Wheelis, M. L. (1990). Towards a natural system of organisms: Proposal for the domains archaea, bacteria, and eucarya. *Proc. Natl. Acad. Sci. USA* **87**, 12, pp. 4576–4579, doi: 10.1073/pnas.87.12.4576, http://www.pnas.org/content/87/12/4576.abstract.

Wolfram (2014). http://www.wolframalpha.com/.

Wolfsberg, T. G., Primakoff, P., Myles, D. G., and White, J. M. (1995). Adam, a novel family of membrane proteins containing a disintegrin and metalloprotease domain: Multipotential functions in cell-cell and cell-matrix interactions. *J. Cell Biol.* **131**, 2, pp. 275–278, doi: 10.1083/jcb.131.2.275, http://jcb.rupress.org/content/131/2/275.short.

Wu, C., Chen, P., Yu, H., Liu, Q., Zong, X., Cai, H., and Wang, P. (2009). A novel biomimetic olfactory-based biosensor for single olfactory sensory neuron monitoring, *Biosens. Bioelectron.* **24**, 5, pp. 1498–1502, doi: http://dx.doi.org/10.1016/j.bios.2008.07.065, http://www.sciencedirect.com/science/article/pii/S095656630800420X, *Selected Papers from the Tenth World Congress on Biosensors Shangai, China, May 14-16, 2008.*

Wu, C., Du, L., Wang, D., Zhao, L., and Wang, P. (2012). A biomimetic olfactory-based biosensor with high efficiency immobilization of molecular detectors, *Biosens. Bioelectron.* **31**, 1, pp. 44–48, doi: http://dx.doi.org/10.1016/j.bios.2011.09.037, http://www.sciencedirect.com/science/article/pii/S0956566311006580.

Wuthrich, K. (1986). *NMR of Proteins and Nucleic Acids* (Wiley, New York, USA).

Yellen, G. (2002). The voltage-gated potassium channels and their relatives, *Nature (London)* **419**, 6902, pp. 35–42.

Yoon, H., Lee, S. H., Kwon, O. S., Song, H. S., Oh, E. H., Park, T. H., and Jang, J. (2009). Polypyrrole nanotubes conjugated with human olfactory receptors: High-performance transducers for FET-type bioelectronic noses. *Angew. Chem. Int. Ed.* **48**, 15, pp. 2755–2758.

Zhao, J., Davis, J. J., Sansom, M. S. P., and Hung, A. (2004). Exploring the electronic and mechanical properties of protein using conducting atomic force microscopy, *J. Am. Chem. Soc.* **126**, 17, pp. 5601–5609, doi: 10.1021/ja039392a, http://pubs.acs.org/doi/abs/10.1021/ja039392a, pMID: 15113232.

Zozulya, S., Echeverri, F., and Nguyen, T. (2001). The human olfactory receptor repertoire, *Genome Biol.* **2**, 6, pp. 1–12, doi: 10.1186/gb-2001-2-6-research0018, http://dx.doi.org/10.1186/gb-2001-2-6-research0018.

Zvyagin, I. P. (2006). *Charge Transport in Disordered Solids with Applications in Electronics*, chap. Charge Transport via Delocalized States in Disordered Materials (John Wiley & Sons), pp. 1–48.

# Index